Safety and health practices
of multinational enterprises

Safety and health practices of multinational enterprises

International Labour Office   Geneva

ISBN 92-2-103742-8

First published 1984

Printed in Switzerland

PREFACE

This report presents the findings of a research project regarding the occupational safety and health standards of multinational enterprises (MNEs) in home and host countries, which was undertaken by the International Labour Office under its multinational enterprises programme (MULTI) in consultation with the ILO's Occupational Safety and Health Branch (SEC HYG). Special emphasis was placed on the question of the transfer of relevant information to the workers in these enterprises, to the pertinent authorities and to workers' and employers' organisations. A particular interest was attached in this context to special or new safety and health hazards, their causes and related protective measures taking the international experience of the multinational enterprise into account.

The matters reviewed corresponded to one of the subjects specifically relevant to areas covered by the Tripartite Declaration of Principles concerning Multinational Enterprises and Social Policy which were proposed by the ILO Governing Body at the recommendation of its Committee on MNEs for study by the Office.[1]

The study covers the operations of eight major MNEs in several industrialised home and host countries (in Europe and the United States) as well as in two principal developing countries where multinational enterprises are located, i.e. Mexico and Nigeria. The enterprises included in the sample fall into three sectors with different types of production processes and hence safety and health risks, namely (i) chemical and related industries such as the petroleum industry; (ii) construction and civil engineering; and (iii) several branches of the mechanical and electronics industries. These are all industries with high technology in which multinationals are primarily concentrated in home and host countries. For practical reasons, other industries of importance in connection with multinational enterprise operations such as textiles and food and drink industries have not been included in the study as they will be covered in two forthcoming sectoral studies on the social and labour practices of multinational enterprises.[2]

The main body of information for this study was obtained through semi-structured, in-depth interviews[3] with the various parties concerned with safety and health standards, namely (i) the management (and safety and health experts) of the headquarters and subsidiaries of the participating enterprises; (ii) workers' representatives in the enterprises and workers' organisations (including information obtained from various International Trade Secretariats); (iii) employers' organisations; and (iv) appropriate governmental authorities and other relevant public institutions, such as labour inspection and accident insurance bodies in the home and host countries covered by the survey.[4] Additionally, a considerable number of relevant sources have been used from the general area of occupational safety and health. The ILO wishes to express its appreciation and thanks to all persons and institutions having contributed in various ways to securing the information base needed for this study.

Although the findings of the study have been derived from a limited sample of enterprises and situations, there is no reason to believe that they do not illustrate well the types of issues found in the area of occupational safety and health and the related way in which information is provided by multinational enterprises. Using this assumption, a number of more general findings are presented in the last chapter. It is hoped that these findings of the study will contribute to an informed discussion of the subject under review.

The study was co-ordinated by Dr. Jerry Purswell, Professor of Industrial Engineering, University of Oklahoma, formerly Chief of the ILO Safety and Health Branch (SEC HYG) who also prepared the present report and conducted interviews in the Netherlands, the United States and in Mexico. The other interviews in Europe and those in Nigeria were undertaken by Mr. John Northcott, former Chief of the Occupational Safety Section in the ILO, who also furnished technical advice for the project. Mr. Paul Bailey, Research Officer, MULTI, was charged with the organisational aspects of the project and assisted with interviews in the European countries and in the preparation of the study report.

Notes

[1] Report in GB.214/6/3, paragraph 85.III adopted by the Governing Body at its 214th Session (November, 1980).

[2] These studies cover (i) the textiles and clothing industries (plus footwear), and (ii) the food and drink industries. Two International Trade Secretariats in these sectors, however, provided information for the present study as shown in Appendix IV.

[3] The indicative interview schedule which was selectively used to fit the various respondents is included in Appendix I.

[4] The list of institutions and/or positions of persons contacted for interviews is found in Appendices II-IV.

CONTENTS

CHAPTER I

INTRODUCTION

World-wide concern for occupational safety and health information

Technological change, the emergence of new products and production processes as well as industrialisation of the developing countries and the role of multi-national in this process has brought the question of occupational safety and health into a new focus. These developments are reflected in a renewed emphasis in recent years on ILO standard-setting in this field and on the related provision of information on safety and health.

In June 1974, the International Labour Conference adopted unanimously a "Resolution concerning the working environment"[1] which requested that "a coherent and integrated programme of ILO action designed to contribute effectively to the improvement of the working environment under all its aspects" be prepared "as a matter of urgency, and in co-operation with other organisations concerned". Two additional resolutions were adopted at the International Labour Conference in 1975 and 1976 dealing with occupational safety and health as well as other working conditions.[2] While many of the 155 member States of the ILO had already taken major steps to improve the occupational safety and health of workers in their respective States, the adoption of the three resolutions by the International Labour Conference served to create an additional stimulus for furthering occupational safety and health conditions in many member States, especially those with an emerging industrial base.

The growing concern of ILO member States and indeed its tripartite constituency as a whole for improved occupational safety and health information and action led to the International Programme for the Improvement of Working Conditions and Environment (PIACT) being launched during the 1976-77 biennium. This programme has served as the focus of ILO efforts to provide a variety of means of action to the member States for improving occupational safety, health and working conditions. [3]The first five years of the Programme (1976-81) were recently evaluated.[3] One of the conclusions of the meeting convened for that purpose was that "... there has been a growing preoccupation with the problem of working conditions and environment in many countries which has led to the inauguration or reinforcement of policies and programme of action defined in relation to the needs of these countries."[4]

As a reflection of the increased interests of ILO member States in the occupational safety and health of their workers, Convention No. 155 and Recommendation No. 164 were adopted by the International Labour Conference in 1981. The Convention is considered a major landmark in calling on all of the member States to set up effective systems to protect the safety and health of the workers in their countries. Of course, one of the most acute needs for many of the ILO member States is adequate knowledge of the occupational safety and health hazards present in the industries within the country. In addition to knowing about the types of hazards present, member States need to know how to effectively organise prevention measures through the implementation of a programme of standards and factory inspection to protect the occupational safety and health of their workers.

One of the major trends in occupational safety and health during the last decade has been an increasing awareness of the threats presented by various materials in the work environment to the long-term health of the exposed workers. Occupational diseases such as those produced by asbestos, with a latency period of 10-20 years from exposure to onset of the disease, have created a new sense of the urgency of understanding the long-term impact of exposure to hazardous materials in the work environment.

MNEs' importance in developing and transferring safety and health information

Multinational enterprises (MNEs) are major employers both in the industrialised and in the developing countries. According to recent ILO studies they directly employed some 44 million workers world-wide by 1980, with 4 million of these workers in the developing countries.[5] It has been noted in these studies that multinational enterprises predominate in industries with a high technology component. These include in particular chemicals, pharmaceuticals, petroleum refining, electrical and other machinery and transport equipment, especially automobiles and spare parts and construction. In recent years there has also been a major upsurge in multinational operations/employment in the computer and information industry as well as in robotisation and automated manufacturing systems. The high technology character of MNEs results in two major impacts on the occupational safety and health of the workers in these enterprises. In the first instance, the high levels of technology applied in the machines and processes used requires that special precautions be taken to protect the workers in the establishments. In some instances the hazards can be adequately guarded by physical barriers to protect the workers. However, in many industries the machines and processes involved require the workers to learn a comprehensive set of procedural steps for their safe operation. This requirement implies that the workers must possess a significant level of understanding regarding the technology involved as well as extensive practice in following the procedural steps necessary for safe operation. This latter requirement become especially important when MNEs set up operations in developing countries.

The second major occupational safety and health problem which may exist in high technology industries relates to worker exposure to a number of materials either associated with the final product or as a part of the manufacturing process for which little may be known about the long-term health effects on the worker of exposure to these materials. A recent example of this problem is the finding that the monomer of vinyl chloride produces cancer when breathed by workers over a period of several years if the time weighted average concentration is greater than 5 parts per million (ppm) in air.[6]

Another aspect of high technology industries which is specific to multi-national enterprises is the possible transfer to other countries of occupational safety and health risks connected with the transfer of operations and new technology, especially to developing countries. In these environments the national provisions for safety and health in the industries in question may indeed not as yet be adapted to such new types of risks. Because of this potential for creating problems in technology-receiving countries, a recent ILO symposium has therefore suggested that "technology-exporting countries should provide all information they have available concerning occupational safety and health to the technology recipient. The technology-receiving countries should provide to the exporter all available information on technical conditions and on economic and social objectives as well as legislation relevant to the particular transfer."[7] This double perspective is relevant mutatis mutandis to the operations of multi-nationals in the various countries. It is clear at the same time that the transfer of technology can provide significant economic benefits for the recipient country - and the workers employed by multinational enterprises - if it is properly applied.   The occupational safety and health problems arise when the country in which the new technology is implanted does not have the necessary controls to ensure that safety and health practices in the enterprises are adequate to protect their workers.   Obviously the MNEs can play a key role in establishing adequate protective measures as the result of the experience they have acquired both in their home country operations where most of the research and the development is located, and also in other host countries, taking into account their human, technical and financial resources.   Where MNEs establish production processes in a country in which the national authorities do not have the necessary experience and/or the technical and economic resources to maintain sufficient standards of protection for the workers, they must apply their own standards based on their international experience so as to protect the safety and health of the workers in their plants, with any adaptations necessitated by the local environment.

Purpose of this study

The major purpose of the present study was to focus on the importance of the MNEs in developing, transferring and implementing adequate safety and health information from their home country operations to operations of their subsidiaries, and to other relevant institutions in the various host countries.

The transfer of information cannot be seen as an intercorporate process only, although intercorporate arrangements and policies allowing systematic learning from the enterprises' overall experience and information flow from headquarters to subsidiaries and feedback from the subsidiaries is a very essential aspect of the overall problem.   The transfer of information on occupational safety and health problems and protective standards is also a matter of communication, exchange and co-operation between MNEs and their workers everywhere, as well as employers' and workers' organisations and host governments, in particular the relevant occupational safety and health organisations.

In some instances the institutional mechanisms for involving workers in safety and health matters are mandated by the national authorities, while in other instances the workers may have bargained with their employer to secure their participation in these matters.   The MNEs may also develop their own procedures for eliciting worker participation in safety and health matters when not required by national authorities or collective agreements.

Recommendations regarding occupational safety and health standards and on related provisions as well as the co-operation of MNEs in the work of international organisations concerned with the preparation and adoption of international safety and health standards, are contained in the ILO's Tripartite Declaration of Principles concerning Multinational Enterprises and Social Policy (paragraphs 36-39).   A certain amount of information on the implementation of these provisions - as well as of all other provisions of the Declaration - is obtained by the ILO through periodic government reports on the effect given to the Declaration.   Two such comprehensive follow-up surveys have been undertaken thus far and considered by the ILO Governing Body Committee on Multinational Enterprises in 1980 and 1983 respectively.

The present study serves a complementary but somewhat different purpose. Its focus is not the application of the standards of the Declaration as such. Its aim is rather an in-depth investigation of the issues involved in the safety and health provisions of the Declaration which, for practical reasons, is only possible for a limited number of actually observed cases (enterprises, countries, products and processes, occupational safety and health risks, etc.). It is expected in this way to foster an understanding of the various problems connected with implementing the provisions of the Declaration and thus to contribute to an informed discussion of the safety and health questions involved, both within the framework of the Declaration and otherwise. Finally, it is expected that this study will provide some practical guidance through its conclusions for the improvement of safety and health standards in MNEs wherever they operate, particularly with regard to the provision of information to all the parties concerned. Even though the sample of enterprises surveyed was limited, it is believed that the study illustrates the range of issues found within all MNEs, particularly as regards training and education and the provision of information.

Methodology for the study

In considering the outcome of any study, there is always a need to understand the methodology followed in order to give proper interpretations to the results. The following sections describe the salient aspects of the methodology used in conducting this study.

Need for in-depth interviews to collect information

Because of the broad scope of occupational safety and health, with the various scientific disciplines that contribute to the field such as occupational medicine, occupational safety engineering, it was decided that in addition to contacts with other parts of ILO's tripartite structure, on-site interviews would be necessary to obtain the full range of information about the safety and health programme of an MNE. The data collection problems were compounded by the different organisational structures found for the delivery of occupational safety and health services by employers, as well as the diversity of the standards and factory inspection services which exist in the various countries. The use of a structured, entirely written questionnaire would no doubt have led to some serious omissions in the collection of information. After consideration, it was decided that an in-depth interview procedure should be followed, with a basic set of questions around which the interviews could be structured (see Appendix I). In many instances it was desirable to pursue the response to a given question until a clear picture emerged regarding the safety and health practices for a product or for an operating unit of the MNE. The questionnaire used to structure the interview reflects the issues covered in the relevant provisions of the Tripartite Declaration of Principles concerning Multinational Enterprises and Social Policy (paragraphs 36 to 39). The thrust of the first group of questions was an understanding of the particular safety and health hazards existing in the enterprise, the standards of the enterprise for regulating these hazards, and the relationship of the headquarters unit to the subsidiary units in the development of standards and procedures for control of occupational safety and health hazards. Of particular importance was the flow of information between the headquarters unit and the subsidiary units regarding occupational safety and health matters. The second major section of the interview was structured to learn the manner in which the MNE informed workers, national authorities and other employers concerning occupational safety and health questions. The last portion of the interview concerned the general co-operation which existed between the national safety and health authorities and the various units of the MNE as well as the relationship of the MNE to international organisations such as the ILO.

Tripartite approach for data collection

Considering the theme and scope of the study, the data collection was organised to obtain information from worker representatives and their organisations, national authorities, employers' organisations and of course the multinational employers selected for the study. (The representatives of management interviewed are listed in Appendix II.) The approach followed for each of these groups will be briefly described.

Worker representatives and organisations at the international, national and enterprise levels were invited to submit information regarding the safety and health practices of multinational enterprises in the related sectors. They were requested to give special consideration to ways and means of facilitating the exchange of safety and health information among workers of a multinational enterprise. A request was made of each multinational employer for interviews with one or more worker representatives at each manufacturing or construction site visited. These interviews were scheduled to follow the visits with the enterprise management in order to have as complete a picture as possible of the safety and health practices of the enterprise as a basis for interpreting the comments of the workers. An important feature during the visits to MNEs and their subsidiaries took the form of random discussions with workers on the shop-floor and at the different sites, as well as with safety stewards and other interested personnel. These discussions resulted in some valuable informal opinions being gained. (Information concerning workers interviewed at each site is also shown in Appendix II. The national or sectorial unions providing information are shown in Appendix III and the international sectorial union organisations in Appendix IV.)

National authorities responsible for occupational safety and health were interviewed in each country where a headquarters or subsidiary unit of a multinational enterprise was located for those partipating in the study. Contacts with the national authorities participating in the study were primarily through the Labour Ministry of each country, although some visits were scheduled with the health ministry of the country as well as with the national social security services in order to obtain a complete picture of the occupational safety and health programmes of the country. A list of the national authorities contributing information to the study can be found in Apppendix III.

Employer representatives were contacted in each country where a headquarters or subsidiary unit of the multinational enterprise participating in the study was located. These contacts with employer associations were of course in addition to the contacts with the mangement of the multinational enterprises participating in the study. Of particular interest was the way in which safety and health information was collected and shared among the employer associations of the country. A listing of the employer associations contributing information to the study is presented in Appendix III.

Timetable of the study

The study was planned during the first half of 1982 and the interviews were completed during the period of June 1982 through June 1983. In each case the headquarters unit of the multinational enterprise was interviewed first, followed by interviews with the subsidiaries participating in the study. It should be noted that a world-wide recession was taking place during the time when the data was being collected. It is believed however that this economic situation did not significantly influence the outcome of the study, although some potential differences exist in the collection of information during a period of economic growth as compared to recession. It is known from a variety of studies that contraction in the workforce during a period of recession generally results in a workforce with more seniority and thus contributes to an improvement in the accident frequency rate of the enterprise because of the more experienced workers. However, the safety and health records of each enterprise were reviewed for a period of time dating back to the period of greater economic activity prior to the recession, and therefore the only potential effect on the study was the slower pace of operations observed during some of the site visits.

Method of collecting and analysing data

The first major concern of the study was the identification of the hazards associated with the manufacturing or construction operations of the multinational enterprise in question. Because the interviewers were knowledgeable regarding the safety and health hazards in a wide range of manufacturing and construction activities, it was possible to begin the data collection at each enterprise with an anticipated list of hazards based on the manufacturing operations expected to be associated with each product. Each enterprise was asked to provide data regarding recent experience with occupational injuries and illnesses and this information in turn was reviewed to add or delete items from the list of hazards for the enterprise. Finally, the safety and health personnel for each enterprise were requested to provide a list of the safety and health hazards they considered to be most important for the enterprise. The data regarding hazards associated with a given type of manufacturing or construction were then analysed in order to serve as a guide for workers, employers, and national authorities in anticipating the type of hazards which workers must be protected against for a given type of industry either in existence or planned for the country. Of particular importance in this data collection and analysis effort were those newly emerging occupational illnesses for which little information has been available until the last few years. Some of the multinational enterprises were found to have very extensive systems for the collection and review of toxicological information associated with the types of materials used in the manufacture of their products.

Given that information is known regarding the hazards associated with a manufacturing or construction operation, the next major data collection effort was directed to appropriate measures for control of these hazards. This step of the data collection and analysis focused on the policies of the enterprise regarding appropriate measures to control these hazards, especially a comparison of the control measures which were employed at the headquarters facilities as compared to the facilities of the subsidiaries participating in the study. The policies and procedures of the enterprise were analysed in relation to the laws and regulations of the national authorities for controlling occupational hazards.

The next phase of the analysis was directed to the manner in which information regarding policies for safety and health was communicated between the headquarters unit and its subsidiaries, and in turn the manner in which the policies were implemented to provide a safe and healthy workplace as well as conform to the requirements of the national authorities.

Finally, an attempt was made to identify those systems for occupational safety and health information exchange which proved to be effective as compared with those systems in which the information exchange for safety and health proved not to be as effective as desirable.

Criteria for selection of MNEs
included in study sample

It has been noted earlier in this report that there is tendency for MNEs to predominate in industries which make relatively intensive use of technology such as chemicals, pharmaceuticals, petroleum refining, construction, electrical and non-electrical machinery and transport equipment, especially automobiles and spare parts. Therefore, any study dealing with the occupational safety and health practices of multinationals should incorporate MNEs which are representative of these industries. Some MNEs specialise in large installations such as power stations, and the construction of facilities related to high technology. It was, therefore, desirable to include in the study sample some of the MNEs which undertake this type of construction.

Ideally, it would have been desirable to study the safety and health operations of an MNE in its home country and then in a number of the host countries where it operates. This proved impossible, however, because of expenditure and time constraints. It was thus decided to incorporate in-depth interviews with the headquarters unit of the MNE selected for the study and then to conduct in-depth interviews in a subsidiary operating in another industrialised country and a developing country where possible. Using such an approach, a comparison could then be made between operations in two industrialised ocuntries, which in turn could further be compared with operations in a developing country.

For practical reasons, two developing countries were chosen where a number of important subsidiaries of multinational enterprises are found. These were Mexico and Nigeria. The chosen industrialised countries where subsidiaries have been established were the Federal Republic of Germany, the Netherlands, the United Kingdom and the United States. The actual choice of the location of the production units was determined in consultation with the management of the MNE headquarters concerned.

MNEs' headquarters and subsidiaries
selected for study

Given the criteria for selection of a sample of MNEs for study, a prospective list of suitable MNEs was developed. Some of the MNEs originally selected for study were unable to participate for various reasons. The MNEs who finally agreed to take part are shown in table 1.

Two MNEs were selected for study within the chemicals and related products industry sector. BASF AG, with headquarters in the Federal Republic of Germany represented the chemicals sector and the subsidiary chosen was BASF (Mexico). The Royal Dutch/Shell group of companies was selected as an MNE whose activities encompassed both petroleum refining and petrochemicals, thus providing an opportunity to collect information from two other sectors of the industry. Shell USA was selected as a subsidiary representing operations in an industrialised country, and Shell-PD Nigeria as a subsidiary operating in a developing country.

Merck, Sharp and Dohme with headquarters (Merck and Company, Inc.) in the United States and a subsidiary in the Netherlands was chosen as a representative multinational in the pharmaceutical sector.

Within the construction industry the Bechtel companies, being an MNE dealing with high technology construction was selected for study; they have headquarters in the United States and a subsidiary in the United Kingdom. This enabled some comparison to be made of the safety and health programmes conducted by a large

Table 1: MNEs selected for study

Sector \ Country	Chemicals	Pharmaceuticals	Petroleum refining and petro-chemicals	Construction	Motor vehicles	Electrical machinery	Office equipment
Switzerland						Brown Boveri (H)	
Federal Republic of Germany	BASF (H)				Volkswagen (H)	Brown Boveri (S)	
Netherlands		Merck, Sharp and Dohme (S)	Shell (H)				
USA		Merck, Sharp and Dohme (H)	Shell (S)	Bechtel (H)			Xerox (H)
Mexico	BASF de Mexico (S)				Volkswagen de Mexico (S)	Brown Boveri (S)	Xerox (S)
United Kingdom				Bechtel (S)		BICC (H)	Xerox (S)
Nigeria			Shell-PD Nigeria (S)	BICC - Balfour Beatty (S)	Volkswagen Nigeria (S)		

H = Headquarters (or central service organisation).
S = Subsidiary (usually understood in the sense of majority owned).

construction company in two industrialised countries. The construction operations of BICC in Nigeria (run by Balfour Beatty) were selected for study to obtain representation from a developing country which could be compared with the operations of Bechtel in the two industrialised countries.

Within the motor vehicles sector, Volkswagen AG in the Federal Republic of Germany was selected as the headquarters unit with subsidiary operations in both Mexico and Nigeria to provide a sample for comparing activities in two developing countries and an industrialised country.

Within the electrical machinery sector, the MNEs included are producers of heavy electrical equipment and components as well as constructors of large electrical installations. Brown Boveri, with headquarters in Switzerland, and subsidiaries in the Federal Republic of Germany and Mexico was chosen to provide a basis for comparing operations in an industrialised country subsidiary with a developing country subsidiary. BICC operations in the United Kingdom also provided further data from an industrialised country. Xerox, with headquarters in the United States, and subsidiary operations both in a developing country (Mexico) and in an industrialised country (United Kingdom) were selected for study in the high technology office equipment sector.

In general, one enterprise from each sector was included, except for construction and electrical machinery, where two additional enterprises were covered.

While the principal focus of the study was on the exchange of information regarding safety and health programmes, it was considered desirable to have some measures of effectiveness for the programmes of the enterprise, such as injury and illness data. In most instances this data was provided by the enterprises as requested. Still the different ranges of products manufactured, the different systems of reporting accidents and sickness, the different cultural, climatic and other factors from one country to another, and, finally, the different structures and systems of safety and health management preclude comparisons in every detail. The study will nevertheless show certain interesting comparisons between one case and another.

Operational and economic characteristics of MNEs included in the study

The purpose of this section is to provide an overview of the economic and operational characteristics of the MNEs studied, based on data collected during the in-depth interviews and extracted from company directories. As relevant, additional information will be introduced in the specific sections dealing with occupational safety and health information related to the MNE.

BASF is headquartered in Ludwigshafen, Federal Republic of Germany. The history of BASF dates back to 1861 when it was founded to produce dyes made from coal tar. The company has traditionally specialised in petrochemicals, industrial chemicals, fertilisers, plastics and dyestuffs, with pharmaceuticals being added to the product mix in recent years. Total sales during the year 1981 reached DM 31,766 million (US$12,217 million). Each of the major product categories mentioned earlier accounted for at least 10 per cent of total sales, with basic petrochemicals and other chemicals accounting for approximately 40 per cent of total sales. A major expansion to manufacturing facilities abroad occurred during the late 1950s. BASF operates in Mexico through a holding company which is wholly owned by the parent company. This company is

known as BASF Mexicana, SA. Through the holding company, BASF has a 40 per
cent ownership in Polioles as well as a 40 per cent ownership in Vitaminas which
are the operating companies. The remaining 60 per cent of each operating company is owned within Mexico. The plant visited in Mexico was located at Santa
Clara, which is a part of the Polioles operating company. It is the oldest of
the BASF operations in Mexico, producing plastic foam (Styropor) for the insulation
of buildings. BASF employes some 116,000 persons world-wide, of whom 75 per
cent are employed in the home country operations,[10] and the remainder abroad. It
operates in some 15 countries in all continents.

Merck and Company, Inc. is a major manufacturer of pharmaceuticals and
animal health products. Other products include chicken breeding stock and
food additives as well as products related to the control of environmental
hazards. Human and animal health products accounted for 77 per cent of sales
in 1981, with the remainder of sales attributed to speciality chemical and
environmental products. Merck and Company is located in Rahway, New Jersey,
the United States. The American pharmaceutical operations of Merck and Company
are conducted under the Merck, Sharp and Dohme Division, and international
pharmaceutical operations are conducted by the Merck, Sharp and Dohme International Division. The subsidiary company visited in the Netherlands was Merck,
Sharp and Dohme BV. Sales of Merck and Company in 1981 were $2,925.5 million,
with 47 per cent of these sales resulting from foreign marketing operations.
Merck operates in some 30 countries in North America, Latin America, Asia,
Africa and Europe and employs approximately[11] 30,000 persons world-wide, of whom
50 per cent are in operations abroad.

The Royal Dutch/Shell Group of companies comprises: two holding companies,
which in their turn own between them, directly or indirectly, shares in all
other Group and associated companies. There are two parent companies: the
Royal Dutch Petroleum Company (Netherlands) having a 60 per cent interest in the
Group, and Shell Transport (UK), 40 per cent. Shell companies operate in more
than 100 countries with ownership of the Shell company in the various countries
varying from less than a majority interest to wholly owned subsidiaries in some
countries. Total earnings from operations in 1981 were £2,811 million
(US$4,216 million). Shell Oil Company-USA operates as a part of the Shell
Group in which the Royal Dutch Petroleum Company and the Shell Transport and
Trading Company, collectively owned (in 1982) a 69 per cent interest. The Shell
Petroleum and Development Company of Nigeria, Ltd. is wholly owned by the Shell
Group. Throughout the world the group[12] employs 166,000 persons, 66 per cent of
whom are outside its two home bases.

The Bechtel Companies include the Bechtel Power Company, Bechtel Civil and
Minerals Company, Bechtel Petroleum Company and the Bechtel Holding Company.
The Bechtel Companies are privately held, so the usual information about public
companies is not available in this instance. The various Bechtel Companies
design and construct large petroleum processing units, nuclear and fossil-fuel
powered plants, and other major installations throughout the world, such as the
entirely new city and airport currently being constructed in Saudi Arabia.
Bechtel Great Britain, Ltd. is located in London and is one of the three
divisions of Bechtel Petroleum, Inc.

The BICC Group is comprised of four group companies; Balfour Beatty, BICC
Cables, BICC Industrial Products and BICC International. Balfour Beatty, Ltd.
is a large contracting organisation which performs engineering and construction
activities on a world-wide basis. BICC Cables, Ltd. manufactures a comprehensive range of electrical wires, cables and conductors, and refines copper for
use in manufacturing these cables. BICC Industrial Products, Ltd. manufactures

and supplies electrical and electronic systems and components for a wide range
of industrial uses.   BICC International, Ltd. is the operating company through
which overseas investments are made in cable making, metals and plastics.
Balfour Beatty, Ltd. and BICC Cables, Ltd. were the two group companies included
in this study in Nigeria and the United Kingdom respectively.   The BICC Group
operates in some 30 countries on all continents.   It employs over 50,000 persons[13]
world-wide, of which some 40 per cent are in foreign operations.   In addition to
discussions with the Chief Medical Officer and his principal safety engineers,
two plants belonging to BICC were visited.   These were the Prescott Works and
the Leigh Works.   These factories are concerned with basic wire production and
cable making, the manufacture of small components and of plastic-insulated conductors.

Volkswagenwerk, AG, located in Wolfsburg, Federal Republic of Germany, is
the largest European car manufacturer.   Volkswagen produces a number of different passenger cars and light trucks.   Major foreign subsidiaries of Volkswagen
are located in Brazil and Mexico.   The subsidiary in Mexico which was included
as a part of this study has operated since 1964.   It is a fully integrated manufacturing facility, with a very modern engine plant as well as manufacturing and
assembly operations for a full range of passenger vehicles and light trucks.
The subsidiary in Nigeria which was also included in this study performs only
assembly operations.   The manufacturing companies abroad such as those in Brazil
and Mexico are operated as separate units with only overall financial control
from the headquarters office.   Volkswagenwerk, AG has a 100 per cent ownership
of Volkswagen de Mexico, while only a 40 per cent share is owned of Volkswagen
of Nigeria, Ltd.   Overall sales of the enterprise for 1981 were DM 37,878 million
(US$14,568 million).   Volkswagen has principal subsidiaries in some ten countries
in Europe, the United States, Latin America and Africa.   It employs close to
250,000 persons throughout the world of whom over 30 per cent are in its foreign
operations.[14]

Xerox Corporation is the leading manufacturer of photocopiers in the world.
The Xerox headquarters in Stamford, Connecticut, USA oversees a world-wide
organisation.   Sales abroad accounted for almost half of the US$8,691 million
of total sales for the year 1981.   The principal products are copiers and
duplicators as well as paper for the copiers.   Xerox de Mexico, SA is operated
as a part of the Xerox Latin-American Group.   The Xerox Corporation has a 75 per
cent ownership in Xerox de Mexico.   The Xerox Corporation owns a 51 per cent
share of Rank Xerox which is the major operating company for Europe.   The Xerox
facilities which were included as a part of the study in the United Kingdom are
operated as Rank Xerox, Ltd.   In addition to its home country operations,
principal Xerox subsidiaries are to be found in more than 40 countries throughout the world.   Its total employees world-wide number more than 100,000.[15]

Brown Boveri and Company, Ltd. is one of the largest electrical engineering
companies in the world.   Its principal products include power generation and
distribution equipment, complete electrical installations for industrial and
transport applications, and a wide range of electrical and electronic equipment.
It is unique among the multinational enterprises selected for the study in that
foreign operations typically account for approximately 90 per cent of total sales
each year.   The parent company is situated in Baden, Switzerland, but the largest
operating group is located in Mannheim, Federal Republic of Germany.   The Swiss
parent company has a 56 per cent ownership of Brown Boveri and Cie, AG
(Mannheim) and a 65 per cent ownership in Brown Boveri Mexicana, SA de CV.
Brown Boveri has principal subsidiaries in more than 20 countries, and it employs
some 95,000 persons world-wide, nearly 80 per cent of whom are in its foreign
operations.[16]

Notes

[1] ILO: "Resolution concerning the working environment", adopted by the International Labour Conference, 58th Session, Geneva, 1974.

[2] ILO: "Resolution concerning future action of the International Labour Organisation in the field of working conditions and environment", adopted by the International Labour Conference, 59th Session, Geneva, 1975 and idem: "Resolution on working conditions and environment", adopted by the International Labour Conference, 60th Session, Geneva, 1976.

[3] ILO: Tripartite Advisory Meeting on the Evaluation of the International Programme for the Improvement of Working Conditions and Environment (PIACT), Geneva, 10-15 February 1982: Review of activities (1976-81) and future orientation, Report by the International Labour Office.

[4] ibid., para. 447, p. 93.

[5] ILO: Employment effects of multinational enterprises in industrialised countries (Geneva, 1981) and idem: Employment effects of multinational enterprises in developing countries (Geneva, 1981).

[6] Occupational Safety and Health Administration, US, General Industry Standards, 29 CFR Part 19100.1000, 1980.

[7] ILO: Occupational safety, health and working conditions and the transfer of technology, Report of the Inter-regional Tripartite Symposium on Occupational Safety, Health and Working Conditions Specifications in relation to Transfer of Technology to the Developing Countries, Geneva, 1981.

[8] ILO: GB/MNE/1980/D.1 and GB.224/MNE/1/1/D.1.

[9] John M. Stopford: The World Directory of Multinational Enterprises (London, MacMillan, 1982).

[10] ibid., pp. 128-130.

[11] ibid., pp. 715-717.

[12] ibid., p. 966.

[13] ibid., pp. 164-166.

[14] ibid., p. 1225.

[15] ibid., pp. 1260-1262.

[16] ibid., pp. 207-210.

CHAPTER II

MAJOR OPERATIONAL CONDITIONS RELATED TO
OCCUPATIONAL SAFETY AND HEALTH

Safety and health legislation/regulations
in headquarters and subsidiary countries

It is not possible in the scope of a study such as this one to fully describe all of the legislation and regulations of the national authorities dealing with occupational safety and health which provide the normative framework
for all enterprises. More detail is to be found in the ILO Legislative Series
in which the major regulations of many of the member States related to occupational safety and health and the working environment are published. The legislative series served as useful references for checking the information which was
collected during interviews with the national authorities in each country regarding their system of occupational safety and health legislation and the programmes
for implementation of the legislation. The highlights from this legislation and
the occupational safety and health programmes of each country where the multinationals participating in this study operate are included in the following
sections.

Switzerland

The basic legislative provisions governing the protection of the worker
against occupational injuries and illnesses are set out in Federal Law No. 3,
of 13 March 1964. Three subdivisions of this law (Ordinances Nos. 1, 2, 3)
deal respectively with general conditions of work, certain special categories
of enterprises and workers, and accident prevention and hygiene in industrial
enterprises. The application of this legislation is under the overall supervision of the Federal Department of Economics through the Federal Office for
Industry, Arts and Crafts (OFIAMT) which comprises a labour protection sub-
division made up of four inspectorates, each covering one of four regions of
the country, together with two occupational health branches in Bern and Zurich.

The practical implementation and enforcement of this legislation is
entrusted to the cantonal authorities each of which is responsible for its own
inspection service, as well as hygiene and fire protection services. These
services vary in size and importance from that of the Canton of Geneva which is
highly developed and well equipped to some of those in the less industrially
developed cantons which may only be of a rudimentary nature because of the type
of economic activity and financial resources available to the cantonal
authorities.

Another important federal law in the field of occupational safety and health
is the Federal Law on Accident and Sickness Insurance of June 1911. Administration of this law is under the supervision of the Federal Department of the
Interior, operating through the Social Insurance Office and the Swiss Accident
Insurance Bureau (SUVA) which is an independent body. Its authorised agencies
include technical inspection groups responsible for high-voltage electricity,
boilers, welding, gas, forestry and two units dealing with accident prevention.
SUVA also collaborates with the engineers' and architects' association and the
machine constructors' association.

A new consolidated law covering all aspects of occupational safety and health has been prepared and which became effective on 1 January 1984.

Plans for the construction of new factories as well as revisions to existing ones, including the processes involved, must be submitted for approval to the cantonal authorities before construction is initiated. These plans are reviewed by the cantonal authorities, who in turn pass them to the Federal Inspectorate for review and approval. In many instances, the plans are also reviewed by the technical inspection services of SUVA before they are finally approved. The plans are then passed back to the cantonal authorities who give the final authorisation for construction.

Under the law, inspectors have the option of stopping any process of manufacture or building construction if they deem that a serious hazard exists. Operation or construction may not be resumed until the hazard has been eliminated.

The cantonal inspectors are required to hear complaints from workers or their representative and to investigate any complaints, including anonymous ones.

Even though SUVA is a private entity, there is a very close relationship with the federal and cantonal authorities in the overall occupational safety and health programme of the country. It is therefore important to review some of the contributions which SUVA makes to the overall safety and health programme. Because the Federal Law on Accident and Sickness Insurance of June 1911 includes a provision of compulsory insurance coverage for each enterprise, SUVA can exercise a significant amount of influence in occupational safety and health matters by requiring certain conditions to be met. SUVA inspectors investigate all serious accidents and have the option to stop any operation which is considered dangerous. They also have access to the injury and illness records of an enterprise and can recommend increases in insurance premiums for firms that have poor records and/or safety and health programmes.

Because of their major role in the occupational safety and health programme of the country, SUVA operates laboratories in the fields of pneumoconiosis, and the health effects of chemicals, ionising radiation, noise and vibration. Limits of exposure are established as MAC (Maximum Allowable Concentration) values based on those of the American Council of Government and Industrial Hygienists (ACGIH) with reference also to the MAK (Maximale Arbeitsplatz-Kozentrationen) values of the Federal Republic of Germany.

SUVA officials noted that the standard of safety and health programmes of the multinational companies operating in Switzerland was above average when compared to those of national firms operating in the country.

Federal Republic of Germany

It will be seen in considering the safety and health programme of the Federal Republic of Germany that there are some similarities to the one in Switzerland, i.e. a major role is played by the private worker compensation insurance system of the country. The overall system of occupational safety and health in the Federal Republic of Germany is complex with many different components.

Two types of legal provisions govern occupational safety and health in industrial workplaces: (1) national statutes (Acts and Ordinances); and (2) accident prevention regulations issued by the Industrial Mutual Accident

Insurance Associations (Berufsgenossenschaften). National statutes cover some of the classical problems of occupational safety such as those relating to steam boilers, lifts and elevators, flammable liquids, explosives and other dangerous goods, as well as ordinances related to protection of young workers and pregnant workers.

Regulations issued by the Berufsgenossenschaften deal with the typical industrial hazards related to machinery such as those found in the manufacture of clothing, ceramics, chemicals and the handling of materials. In general, the regulations of the Berufsgenossenschaften complement the national statutes to produce complete coverage of occupational safety and health problems. National Acts are sometimes supplemented by one or more ordinances which contain detailed instructions on how to comply with the safety conditions set forth in the Acts. Accident prevention regulations are also accompanied by rules for their application which indicate a means of complying with the technical requirements contained in the regulations. General implementing ordinances and rules for application do not have the force of law, but nevertheless exert a strong influence over the safety and health conditions found in the workplace because they are used as guide-lines by the inspection services to evaluate compliance with Acts or regulations.

Other important sources of information for occupational safety and health in the workplace are the DIN (Deutsche Industrie Norm) standards and the VDE (Verband Deutscher Elektrotechniker) regulations, directives, etc. These standards and regulations provide detailed guidance for achieving safety and health in the workplace and they are often referenced in national ordinances, and other safety rules and regulations (see figure 1).

Figure 1: Schematic showing the relationships between the various laws, codes, ordinances and rules governing occupational safety and health in the Federal Republic of Germany

Laws	Reichs Insurance Code
Ordinances	Accident prevention regulations
General implementing ordinances	Rules for application of the accident prevention regulation
Generally acknowledged safety engineering and occupational health rules - DIN standards - VDE regulations - directives, etc.	Generally acknowledged safety engineering and occupational health rules - DIN standards - VDE regulations - directives, etc.

Responsibility for occupational safety and health rests with the employer, but the law requires that advice, and in some instances agreement, must be obtained from various parties regarding safety and health conditions in the workplace. Under the law (the Occupational Safety Act) the employer is required to appoint plant physicians and safety officers according to the degree of hazard present in the industry as well as the size of the plant. In addition, all plants employing more than 20 workers must appoint safety stewards, who may, on their own initiative, make recommendations or suggestions within their own assigned area of responsibility. They do not fall into the hierarchical supervisory line and cannot give direct instructions, which remain within the power of the normal supervisors who alone can take any initiative necessary for their implementation. The plant physicians and safety officers work closely with the safety stewards in the performance of their duties to provide advice and information.

If the factory is of a sufficient size to require a plant physician or safety officer, then the employer is also required to appoint a safety and health committee which includes the following members:

- the employer or his representative;

- two members of the works council ("Betriebsrat") nominated by the council;

- the plant physician(s);

- the safety officer(s);

- three safety stewards.

The safety and health committee consults on matters of occupational safety and health and accident prevention at least once each quarter.

In smaller plants without a plant physician or safety officer a safety committee must be set up when more than three safety stewards have been appointed. The employer must confer with the safety stewards or the safety committee at least once a month with participation of the "Betriebsrat".

It should be noted that the works council is selected by the workers and is part of the industrial co-determination system within the Federal Republic of Germany. In its role as representative of the workers, the works council generally acts as watchdog over compliance with the laws and regulations affecting workers, including occupational safety and health matters. Because the employer must consult with the works council and obtain their endorsement when planning new facilities or a revision of existing facilities, the works council has an important role to play in occupational safety and health. Any changes in workplaces, working methods and processes, technical plant, and building facilities will be reviewed by the works council to see that the workers' safety and health are adequately considered.

The labour inspection function is performed jointly by the constituent states of the Federal Republic (Länder) and the "Berufsgenossenschaften". Labour inspectorates of the federal states are organised on a regional basis, where all establishments registered in the district are inspected on a regular basis. On the other hand, the inspectors of the Berufsgenossenschaften are organised according to the various branches of industry, rather than regionally. Thus, the approximately 2,300 state inspectors and the approximately 1,000 industrial Berufsgenossenschaften inspectors complement one another in their relative inspection functions. The inspectors from both the federal states and the Berufsgenossenschaften are required to consult with the works council on the occasion of plant visits in order to exchange information with regard to the safety and health conditions in the enterprise.

If firms in the Federal Republic of Germany send employees abroad under contract, the same accident insurance regulations are applied for their protection as if they were working at home.

A description of the occupational safety and health system would not be complete without mention of the Technical Equipment Act - also known as the Guarding of Machinery Act ("Maschinenschutzgesetz"). This Act requires both manufacturers and importers of equipment to introduce onto the market only machinery and equipment that complies with the occupational safety and health regulations and the accepted rules of good engineering practice. Thus, the responsibility for safety and health in the workplace is broadened to include the manufacturer or importer of equipment, rather than leaving the employer to attempt to determine whether the equipment meets the required standards.

In order to implement the Act it is necessary that standards be established for the testing of equipment and that recognised centres be available for the testing of equipment. Approximately 50 such centres now exist in the Federal Republic of Germany, each with its recognised field of competence. Equipment which meets the test criteria is given a distinctive symbol indicating that it has met the required standards, thus facilitating the examination of such equipment by the labour inspectorate when viewed in the factory. While it requires a significant investment and the establishment of testing facilities and cost of equipment testing, such a programme would appear to have major advantages in implementing an overall occupational safety and health programme within a country.

This brief description of the occupational safety and health programmes within the Federal Republic of Germany will hopefully give the reader an appreciation for the manner in which the different components of the overall system interact to achieve the goal of protecting the safety and health of the worker.

United Kingdom

The legal framework and the specific requirements relating to occupational safety and health are contained in the Health and Safety at Work (HSW), etc. Act of 1974. The Act is the latest of many Acts dealing with health and safety which have been enacted in the United Kingdom, beginning in the last century.

The Act provides for the establishment of the Health and Safety Commission and the Health and Safety Executive and for their constitution and functions as well as for their powers. These powers include the enactment of regulations and the approval of codes of practice relating to safety and health. The Act also provides for the Executive to make arrangements for the enforcement of the relevant statutory provisions and for the appointment, powers and responsibilities of inspectors. A major innovation introduced within the HSW Act is the requirement that employers organise their health and safety activities, committing these arrangements to paper in a document which is open to inspection by the employees and competent authority. This policy document is considered very important by both the Executive as well as the management of the enterprise because it sets goals for occupational safety and health and provides a basis to monitor progress in achieving these goals. The Health and Safety Commission and the Executive is organised into a number of major areas as follows:

(1) Operational branches for factory inspection:

> mines and quarries;
> factories;
> explosives;
> alkali and clean air, hazardous installations;
> nuclear installations.

(2) Safety policy.

(3) Hazardous substance policy.

(4) Resources and planning.

(5) Medical services.

(6) Research and laboratory services.

(7) An Accident Prevention Advisory Unit (APAU) has been in existence since the 1960s before the HSW Act came into being. It has no enforcement role and exists purely as an investigatory and advisory group. It is working at the present time primarily on what are described as studies of safety and health at the major enterprises throughout the United Kingdom. The Unit is currently staffed with six full-time inspectors who have the option of drawing upon the expertise of specialists in various fields as necessary to conduct the in-depth investigations in which they are presently engaged. The reports which result from the investigations are reviewed by the Executive and are confidential to the company concerned. Studies are initiated only upon the request or agreement of the company involved. The Unit is currently embarking on an in-depth study of BICC, Ltd., one of the enterprises included in this study. Because the enteprises selected for study are often some of the largest in the United Kingdom, they also may be multinational enterprises as well. Some of the relevant findings of the Unit are referenced in latter sections of this study.

Factory inspectors in the United Kingdom exercise considerable judgement in applying the standards when making their inspections as compared to the Federal Republic of Germany. In general, it was recognised that the particular method chosen for dealing with a hazard will result from joint discussions between the factory inspector, the employer and the workers' union. Factory inspectors are expected to be able to furnish advice on how best to eliminate a particular hazard and ensure compliance with the standard, while at the same time permitting other equally effective options to be utilised by the employer. In the event of failure to achieve agreement with the employer concerning elimination of a hazard, the inspector has the option to take the employer to court to obtain compliance.

United States

The Occupational Safety and Health (OSH) Act of 1970 is the basic piece of legislation governing the occupational safety and health programme in the United States. This Act includes the following major provisions:

(1) The establishment of an Assistant Secretary of Labor for Occupational Safety and Health Administration (OSHA) who is empowered to establish standards for the control of occupational safety and health hazards in the workplace and to implement a system of factory inspection to assure compliance with the standards.

(2) A National Institute for Occupational Safety and Health (NIOSH) was created under the Secretary of Health and Human Services to perform research regarding occupational safety and health and make recommendations to the Assistant Secretary of OSHA regarding necessary standards.

(3) Both OSHA and NIOSH were assigned responsibilities for the training necessary to implement the Act.

(4) Federal control was established over all factory inspection activities, with the option of delegating these responsibilities to the individual states if the state programmes could be demonstrated as equal in effectiveness to the federal inspection programme.

(5) A system of penalties was established for non-compliance with the standards established by OSHA, and a legal review system was established for hearing cases where the penalties assigned were contested by the employer.

The OSH Act required the Occupational Safety and Health Administration to examine a large body of nationally recognised consensus standards which had been developed as a co-operative effort of industry prior to the Act, giving these voluntary standards the effect of law within a period of two years after passage of the Act in 1970. This provision of the Act has resulted in a large body of standards being enacted which set many detailed requirements for compliance by employers. These standards were given the force of law without the usual lengthy process required to establish regulatory law in United States, which can sometimes continue for a period of three or four years before a new regulation becomes mandatory for compliance by employers. As a consequence of the incorporation of a large volume of recommended standards into the body of required standards initially, as well as the lengthy process required for establishing standards after 1972, the typical inspector finds that many detailed rules must be enforced, while control of other recognised standards must wait for completion of the rule-making activity. In the absence of a specific rule for dealing with a hazard, the factory inspector can require abatement of the hazard under a provision of the Act which sets out the general responsibility of the employer to provide a safe and healthful work environment. However, the factory inspector is called upon to produce a significant amount of supporting information for this type of citation as compared to those where an existing standard has been clearly violated.

The OSHA factory inspectors have the legal right of entry into any establishment for purposes of completing an inspection of the premises. Some of the inspections result from anonymous complaints whereas others are in response to the injury record of the industrial sector of which the firm is a part. Inspectors request the participation of worker representatives during the inspection process, and in most instances the worker is paid by either his employer or union while participating in the inspection.

As compared to the Federal Republic of Germany, there are no requirements in the United States for machines to be tested and certified that they meet required safety standards prior to their sale to an enterprise.

Approximately one-half of the 50 states comprising the United States now operate their own inspection programmes under the auspices of the Federal OSHA. A variety of insurance schemes exist across the 50 states for providing compensation to workers who are injured as well as payment of medical treatment expenses. Although there are exceptions, it is generally believed by employers that the change in premiums paid by an individual enterprise as a result of its injury and illness experience is not sufficient motivation for most establishments to improve their safety and health records.

Individuals, including workers in the United States have the right to bring a civil suit against the manufacturer of any item of equipment or system in which they are injured. This right is frequently exercised by workers, approximately half of whom win monetary awards which are significantly greater than those which they may also collect under the state worker compensation programmes. Thus, while there is no legal requirement that machines be tested and approved prior to their sale, the prudent manufacturer will carefully review the safety and health aspects of his machine before selling it in order to avoid costly lawsuits at a later date. Since almost all manufacturers purchase liability insurance against the prospect of future lawsuits, a significant influence is exerted by the liability insurance inspectors to require the machine manufacturer to meet all applicable safety and health codes, whether they be government mandated ones or voluntary ones established within a given industry.

The Mine Safety and Health Act of 1969 created the position of Assistant Secretary of Labor for Mine Safety and Health Administration (MSHA) to oversee the safety and health of workers in mining and milling operations. MSHA has powers similar to those of OSHA to establish standards and enforce the standards through the assessment of penalties in order to protect the workers in mining and milling operations.

Nigeria

The Factory Inspectorate is a branch of the Ministry of Employment, Labour and Productivity (MELP). The legal basis for the Inspectorate and related enforcement procedures derive from the Factories Act of 1955. At the time of the interviews in 1983, a draft was being prepared to revise the Factories Act. Legislation in other countries such as the United Kingdom as well as international standards such as ILO Conventions and Codes of Practice were considered in the preparation of the draft. One important feature of the proposed Act will be provisions relating to medical inspection of workers and workplaces as well as the appointment of factory physicians and safety specialists. Responsibility for the occupational health provisions of the new Act will rest with the MELP whereas they have previously been under the administration of the Health Ministry.

The factory inspectorate has its headquarters in Lagos, with 19 offices spread throughout the country, each office being responsible to the federal authority but under the administrative control of the province where it is located. In their enforcement role, inspectors are expected to use their initiative and exercise significant judgement in establishing measures for the prevention of accidents and illnesses. Some concern was expressed that the inspectors could provide opinions regarding protective measures which could later be the subject of legal action against them in the event of a subsequent accident. The factory inspectorate is currently hampered by the lack of both material resources and staff, with transport problems making visits to outlying factories difficult. A continuing problem exists in the training and retention of qualified factory inspectors, since those achieving the highest levels of competence are typically recruited by industry with the offer of significantly better pay.

Monthly reports on accidents are prepared and sent to the headquarters office, together with statistical returns, which are later consolidated into an annual report for publication by the MELP. Difficulties were noted in achieving a timely publication of this annual report, since the last annual report printed was for the year 1978. Other safety and health statistics are produced by the National Industrial Safety Council, which is financed largely by the

federal Government.   Problems apparently existed in this organisation as well
in producing timely reports.

The Netherlands

Major changes were occurring in the safety and health legislation of the
Netherlands during the course of the study.   The Working Environment Act which
was passed on 8 November 1980 became effective on 1 January 1983.   Prior to this
time, the principal statute dealing with occupational safety and health was the
Industrial Safety Act of 1934.   Various decrees were passed under the Industrial
Act of 1934 relating to occupational safety and health as well as a number of
Acts dealing with specific hazards such as silicosis or work under high atmos-
pheric pressure.

A factory inspectorate system operates under the Minister of Social Affairs
to enforce the various Royal Decrees dealing with occupational safety and health
as well as specific provisions of the various Acts dealing with occupational
safety and health.   Prior to the new Work Environment Act taking effect in 1983,
the factory inspectors could impose only limited sanctions for failure of an
employer to meet the required provisions of decrees of statutes dealing with occupational safety and health.   As a consequence of the Working Environment Act,
the factory inspector will be able to impose significantly larger penalties for
failure of an employer to meet acceptable safety and health standards.   The
factory inspector is required to evaluate any economic advantage gained by an
employer in failing to comply with a particular requirement and then to assess a
penalty which will offset this economic advantage as well as imposing an additional penalty in an amount up to 100 per cent of the calculated penalty based
on the economic advantage gained.   While the system had not been implemented at
the time of interviews, it would appear that such a system would provide a very
strong incentive for the employer to comply with all of the applicable regulations.

In a number of different interviews, it was noted by the representatives of
both the workers and the employers that the Working Environment Act emphasised
the need for workers and their employer to work jointly in solving the occupational safety and health problems of the enterprise.   A major section of the
Work Environment Act deals with co-operation and consultation between workers and
their employer through the vehicle of a working environment committee.   The Act
establishes specific requirements for the selection and certification of the
working environment committees and gives them broad powers to deal with questions
of safety and health in the work environment.   The intent is that the working
environment committee serve as a vehicle for finding solutions to safety and
health problems which can be agreed upon by both the workers and their employers.
However, the Act includes a provision that permits the worker representatives of
a committee to require that the factory inspectorate enforce the relevant regulations if an agreement cannot be reached with the employer about how they should
be applied in the factory.

The obligations of employers, workers and the factory inspectorate are
clearly stated in the Working Environment Act.   It is interesting that a major
emphasis is placed on the development of a safety and health policy by the enterprise and the implementation of this policy.   These provisions of the law are
similar to the comparable provisions of the Health and Safety at Work Act in the
United Kingdom.

It was noted by the factory inspectorate that little useful information had been collected over time regarding the incidence of occupational disease among workers. Because all workers are compensated when they experience an injury or illness, without having to prove that the injury or illness was a consequence of their employment, it had been considered that no need existed to compile such statistics. Under the new Act, such statistics are required to be maintained by employers. In addition, the new law requires that certain groups of workers be given medical examinations prior to being assigned to a particular job as well as the provision that the Inspectorate of Labour may require medical examinations for certain groups of workers if it is considered that the examination is in the interest of the health of the workers concerned.

It is clear that full implementation of the Working Environment Act will result in many changes in the occupational safety and health system of the Netherlands. In general, the factory inspectorate will promulgate and enforce many more regulations than in the past, and have available much more stringent penalties for failure to comply with these regulations. However, the emphasis of the Act is on employing industrial democracy through the vehicle of the working environment committees to solve the safety and health problems of the enterprise.

Mexico

Legislation governing occupational safety and health in Mexico derives from Article 123 of the Constitution. The Federal Labour Act of 2 December 1969, with amendments in 1978 and 1980 serves as the statutory basis for the regulation of occupational safety and health in the country. The Federal Labour Act of 1969, paragraph 527, defines a number of major industries for which the federal authorities are responsible for the application of labour standards, including occupational safety and health standards. These industries include petrochemicals, mining, steel, electrical and textile products, and a number of other industries usually considered the basic industries of a country. All industries not specifically named as federal responsiblity are considered to be the responsibility of the individual states for enforcement of occupational safety and health standards.

The Federal Labour Inspectorate is a part of the Ministry directed by the Secretary of Labour and Social Services. Approximately 230 federal inspectors are responsible for enforcing safety and health regulations within the federal industries.

The Mexican Institute of Social Security and the Ministry of Health are both involved in the overall occupational safety and health programme of the country. The Institute of Social Security is supported in equal parts by the contributions of workers, their employers and the Government. An extensive system of hospitals and clinics is operated throughout the country to provide medical services to the general population, including the workers who may suffer from injuries or illnesses related to their occupation. Therefore, the Institute of Social Security has a major interest in reducing injuries and illnesses in the workplace in order to control the costs for medical services and the payment of lost wages for injured workers. The Institute offers a consultation service for employers to answer questions related to occupational safety and health. In addition, accidents are investigated and data compiled to assist employers in developing programmes to aid employers in reducing injuries and illnesses in the workplace. The various industrial enterprises in the country are divided into five groups according to the risk or injury or illness to the workers in that classification, and classification determines employer payments to the Institute. The employers thus have an incentive to improve their safety and health programmes in order to reduce the incidence of injury and illness and in turn recieve a reduction in the contribution which must be paid for the support of the social security services.

The Federal Labour Act of 1969 sets forth a very extensive list of injuries and illnesses which are considered to be occupationally related. In addition, extensive tables exist which give a range for the percentage of disability which is to be awarded a worker as a result of an occupational injury or illness. The Ministry of Health is required to certify the occupational physicians who in turn make the judgements regarding the percentage of disability of the worker on which his compensation is based.

The multinational management interviewed indicated that a close working relationship exists between the in-plant, first aid or medical services and the social security services. In most instances, the employers interviewed found that they had more contacts with the Institute in terms of consultation regarding accident prevention than they had with the federal labour inspectors. Federal labour regulations require that a safety and health committee consisting of equal representation of workers and employer members (Comision Mixta) be organised at each enterprise. This committee must meet once each month, make an inspection of the facilities and prepare a written report of the meeting. A copy of this report must be filed with the Secretary of Labour and Social Services.

It was noted that the low pay scale of the federal inspectors and of the social security accident prevention specialists made it difficult to retain highly qualified people in these functions.

Major safety and health hazards related to operations of MNEs participating in the study

BASF AG

The hazards associated with operations in the BASF facilities visited are typical of those associated with any enterprise in the chemical industry. There is a hazard of fire and explosion associated with many of the materials which are processed. The prevention of fires and explosions begins with the original design of the processing units and continues through the development of special operating procedures which the workers must follow. Finally, each chemical plant typically has a well-trained fire brigade to extinguish small fires before they become major ones. In this connection, an explosion killing six workers took place in a factory manufacturing chemical products at the BASF subsidiary BASF Mexicana SA/Santa Clara in 1981.

Workers may experience both acute and long-term health effects when they are exposed to chemical substances being processed. The manufacture of inorganic pigments for paints can be a potential source of hazardous materials such as the various compounds of lead, which can be breathed by workers and thus create lead poisoning. In general, the approach to controlling this type of worker exposure is the use of engineering controls whereby the equipment is designed to contain the hazardous materials within the process and not permit them to be released into the workplace. As a secondary measure, when engineering controls are not utilised, workers should be carefully fitted with respirators which do not permit the entry of the materials into the respiratory tract.

Any operation where workers must move materials about, work on surfaces that can sometimes become slippery, and climb ladders - represent hazards, all of which are present in the chemicals industry - which can be expected to produce injuries unless proper measures are taken. Injuries resulting from slips and falls were reported to be one of the most important sources of injury for the BASF facilities in Ludwigshafen.

A common thread throughout all of the interviews conducted for the study was the problem of injuries produced by traffic accidents where workers in the course of their employment must operate motor vehicles. In addition, the occurrence of a motor vehicle injury while a worker is proceeding directly to and from his workplace is also considered to be an occupational injury in some countries thus creating a bias in the normal injury statistics expected for a factory. Motor vehicle accidents are considered to be an especially significant source of injury to workers in the developing countries because of the lesser developed state of the traffic system and the inexperience of some of the drivers.

Merck and Company, Inc.

The manufacture of pharmaceuticals presents some unique occupational hazards. By definition, the pharmaceuticals being manufactured are expected to produce a pharmacological effect on the human or animal for which the drug is intended. Therefore, the workers who are exposed will experience the effects of the drug if it is ingested into the body in a form which will permit it to enter the bloodstream. The most common form of this type of hazard is presented by dusts which are produced as a result of the processing of drugs into tablets or capsules. Merck management believes that worker exposure to these dusts should be controlled by engineering design of the processing equipment to contain the dust within the equipment to the extent that is feasible. When this proves to be impractical, appropriate respirators are provided and fitted to the workers to prohibit the entry of the dust into the body.

Because of the special precautions taken to control worker exposure to the drugs being manufactured, other types of hazards produce more worker injuries, such as slipping and falling, which is the most common source of injury.

A number of the raw materials as well as some of the finished products being processed can be flammable or explosive, especially dusts which can be disbursed into the air and ignited. These hazards are controlled through careful engineering process design and procedures which must be followed by the workers in the process.

The development and testing of drugs often requires that a significant animal population be available for this purpose. Since workers must care for and perform tests on these animals, the possibility exists for worker injuries due to bites and scratches as well as workers becoming infected with some disease with which the animals have been deliberately or inadvertently infected.

The development of vaccines presents problems of worker exposure to the viral or bacterial agent which must be carefully monitored and controlled. For example, Merck was developing a vaccine for hepatitis, thus presenting the possibility of exposing workers to this virus unless proper safeguards were employed.

In addition to the hazards unique to the pharmaceuticals being processed, there is also the usual range of hazards from operating complex equipment of both a chemicals processing and mechanical nature such as reactors, tablet-making machines, conveyors and forklift trucks.

Royal Dutch/Shell Group

Petroleum refining involves the processing of various types of hydrocarbons, all of which have a degree of flammability and many of which can form highly explosive mixtures in the air. It is necessary that careful consideration be given in the design of petroleum refinery processing units to the prevention of fires and explosions. Similar care must be exercised in the design and operation of petrochemical processing units.

The compounds present in petroleum refining and petrochemical manufacture can produce both acute and long-term health effects. The most common acute effect is asphyxiation as a result of breathing a gaseous hydrocarbon such as methane or ethane. Benzene is present as a part of many petroleum refining and petrochemical operations, and has been linked to the occurrence of leukaemia in human beings. A number of countries are currently reviewing their existing exposure levels for benzene in light of developments in the United States. An exposure standard of 10 ppm in air of the Occupational Safety and Health Administration of the United States is now being reviewed as a result of a recommendation in 1976 by the National Institute of Occupational Safety and Health (United States) to lower the exposure level "as low as possible".

In common with most processing industries, the operation and maintenance of processing units can expose workers to a variety of slipping, tripping and falling hazards. Workers often perform maintenance in confined spaces in which they can be exposed to decreased levels of oxygen. Asphyxiants may leak into the confined space without the workers' knowledge. General practice, however, is to monitor the confined space for the absence of asphyxiants or other toxic materials and the presence of sufficient oxygen.

The exploration, drilling and production operations necessary to obtain petroleum from the earth can expose workers to a wide variety of hazards. In addition to the fire and explosion hazards previously noted, and because of the widely distributed operations involving the use of motor transport, accidents with moving vehicles in traffic or when moving heavy equipment are a frequent occurrence. Working or being transported over water also presents the hazard of drowning.

The Bechtel Companies

Construction operations typically involve a higher incidence of injury then general manufacturing operations because of the new hazards which are introduced during each phase of the changing workplace as the construction progresses. As might be expected, slips, trips and falls, especially falls from heights consti-tute one of the important sources of injury to construction workers. Eye injuries are frequently produced by flying particles of all types which are pro-duced through working on materials at the construction site, especially with workers assuming a wide variety of positions in order to perform their tasks.

The handling of heavy objects has the potential for injuring workers when they are caught between a heavy object and part of a structure. Workers must use a wide variety of powered machines which have the potential for producing serious lacerations, contusions and puncture wounds.

Construction workers are often exposed to suffocation as a result of working in confined spaces, as well as in trenches, which can cave in upon them unless they are properly shored.

The ever increasing use of ionising radiation sources for the checking of welds in metal structures can expose workers to overdoses of ionising radiation. The use of various types of welding machinery for different materials can expose workers to harmful fumes.

The wide variety of adhesives and coatings now used in construction can produce harmful health effects if they are not used properly by the workers.

The BICC Group

The hazards associated with BICC Cables, Ltd. are those which might be expected in the production of a large variety of wire and cable which can be insulated with a number of different coatings. In the case of the basic wire production, the hazards encountered are the usual mechanical hazards of wire-drawing, rotating machinery, snapping and whipping of the wires and the transport of heavy reels of wire. Chemical, metal and solvent fumes present in the processing of the wire can also present a variety of hazards to workers, such as burns, acute and chronic respiratory problems and dermatitis.

The compounding of a variety of plastics with various colouring materials for applying the insulation to the wires also can present a number of hazards to the workers. In most instances, the appropriate method of dealing with the hazards is to enclose the processes as much as possible to prevent exposure of the workers to either the mechanical or chemical hazards present.

In the case of Balfour Beatty, Ltd., the hazards present are those related to construction operations which have been described earlier for the Bechtel Companies. In addition, it was noted in Nigeria that malaria is a problem in some parts of the country as well as bites from poisonous snakes or other animals. Some problems have been encountered from dehydration when working in trenches or other excavations in the high prevailing temperatures.

Volkswagenwerk, AG

The manufacture of automotive vehicles presents a wide range of hazards to workers. The production of vehicle parts in foundries can expose workers to burns, respiratory problems as a result of dust inhalation, eye injuries from foreign particles such as dust and metal and vibration-related injuries resulting from the use of chipping hammers for cleaning castings after metal pouring.

The use of a large variety of presses for forming the metal shapes used on the vehicle, present a variety of hazards to the workers, the most serious being the amputation of fingers and hands in the dies of the press, to lacerations resulting from handling parts with sharp metal edges. The use of many types of metal cutting machines to produce vehicle parts gives rise to the hazard of entanglement of the worker or his clothes in the machine, which can result in

serious injury.   The large variety of conveyors used for moving parts throughout the facility also offers the possibility that workers can become entangled
in the conveyor and seriously injured.

The welding operations necessary to join parts of the vehicle can produce
harmful fumes and in turn create a hazard for the worker unless they are properly
ventilated to the outside of the building.

As the use of robots becomes more common, there is the potential for injury
to production and maintenance workers who may inadvertently come in contact with
these machines during a part of their cycle.

The variety of paints and solvents used in finishing the vehicle present a
number of fire and explosion hazards as well as health effects if they are
breathed in sufficient concentrations by the workers.   Epoxy compounds, which
are used for bonding operations, have the potential for adversely affecting the
skin, mucous membranes, lungs, central nervous system and the liver if not
properly used in the workplace.

Xerox Corporation

Xerox produces a wide range of copy machines which consist of electrical,
electronic and mechanical components.   Production of the mechanical components
requires workers to operate a variety of metal forming machines in which they
can become injured unless proper precautions can be taken.   The operating points
of the machines must be guarded to prevent the entry of the worker's body into
the danger zone.   In those instances where the machine cannot be adequately
guarded, the workers must be carefully trained and instructed to avoid the danger
zones.   Workers may be exposed to various solvents during the cleaning of machine
parts, as well as to fumes which are produced during welding and soldering operations.   Other hazards exist in the cleaning and painting of the metal cabinets
for the machines, requiring localised ventilation and also the wearing of
respirators in some operations.

In addition to the typical hazards mentioned above, some unusual hazards
are present in the manufacture of copy machines.   The material used to produce
the photo image during the copying process is a mixture of arsenic and selenium
over a nickel or aluminum base.   Worker exposure to these materials must be
controlled through the use of adequate localised ventilation, and personal protective equipment.

Another unusual material to which the workers are exposed is a very fine
powder called "toner" that produces the photocopy image in the machine.   The
toner consists of carbon black as well as a variety of plastics.   The carbon
black and the plastics are combined and then ground to a very fine powder.
Inhalation of the dust produced during processing of the toner can present a
hazard to workers.   It was found some years ago that toner produced mutogenic
effects when subjected to toxicological testing.   Subsequent tests showed that
the carbon black was contaminated with trace amounts of nitro-pyrenes, which had
produced the mutagenic findings.   Xerox established stricter quality control
limits for assuring the purity of the carbon black being used in the process and
eliminated the problem.

A significant proportion of the injuries to workers involved chronic musculo-
skeletal problems, lacerations, and puncture wounds from splinters.   Many of the
injuries resulted from manual handling of materials.   The problem of safety

involving the use of motor vehicles was considered to be a major one because of the very large number of Xerox technicians travelling each day to service machines throughout the world. A comprehensive fleet safety programme is provided for these employees to reduce the number of injuries resulting from motor vehicle collisions.

While it was not viewed as a direct problem for Xerox employees, the safety and health staff noted that Xerox manufactures office automation equipment and was therefore interested in all aspects of the safety and health of workers using these machines. Xerox is currently sponsoring research studies and collecting information on a world-wide basis concerning the safety and health problems of workers using office automation equipment.

Brown Boveri and Company, Ltd.

Much of the equipment produced by Brown Boveri involves the fabrication and handling of large parts which are components of machines or electrical power transmission and utilisation equipment. The possibilities of workers injuring themselves in lifting or manoeuvering these parts is one of the recognised hazards in the enterprise. Slips and falls were also considered to be a problem when working on and around the large pieces of equipment. The problem of worker contact with PCBs (polychlorinated biphenyls) is also a problem in servicing some of the older equipment, whereas the newer equipment uses a substitute for PCB which is believed to be less toxic to man, although it does present a greater fire hazard.

The large machines involved in processing the parts of the different products can present hazards to workers if they are not properly guarded or interlocked. It was noted that operators were well protected in the normal use of the machine, whereas some maintenance workers were injured because of difficulties in servicing the machines. Even though a large part of the product involves electrical apparatus, very few accidents are related to electrical shock, but those which do occur are frequently fatal.

It was noted that two hazards which had existed in prior years have been eliminated, one involving the use of asbestos and the other the use of sand containing silica in the metal cleaning operation of sand blasting.

Safety and health programmes of MNEs participating in the study

The organisation and operation of the safety and health programmes of the MNEs participating in the study will be presented in the following sections. As noted earlier, information was obtained from a large number of people in the various multinational enterprises as shown in Appendix II.

BASF AG

The safety and health organisation can be described as one which exercises centralised control in the establishment of policies and their implementation for the protection of their workers' safety and health in all BASF plants throughout the world. It is the intent of BASF management that the safety and health policies be uniformly followed in subsidiaries throughout the world, even in those subsidiaries in which BASF has less than a 50 per cent ownership. In

accordance with the statutes described earlier for the Federal Republic of
Germany, BASF appoints plant physicians and safety engineers in all of the
operations within the Federal Republic of Germany. Plant physicians and safety
engineers are also appointed in subsidiary operations such as the one visited in
the Republic of Mexico. The functions of occupational medicine and health protection as well as safety and plant security are co-ordinated throughout the
enterprise at the headquarters in Ludwigshafen. It is noteworthy that BASF
appointed the first plant physician in 1866, continuing through eight generations
of plant physicians up to the present time.

Extensive written policies and procedures are provided for implementing the
safety and health programmes in the various plants of the enterprise. Because
of the high importance attached to the prevention of fires and explosions, the
headquarters unit of the fire prevention and security service as well as the
technical safety and health staff department regularly visit BASF subsidiaries
to review procedures and provide training and advice. It was noted that many
sources are considered in developing the control limits for various chemical
substances as well as the safety procedures to be followed. Among the organisations consulted for information were the WHO, ILO, EEC, Council of Europe,
Medichem, IRTPC (cancer research), as also are all their relevant publications.

Each new worker receives instructions regarding safety and health practices
of the enterprise and attends a two-and-a-half day course on safety procedures
before beginning his work. Workers are monitored in the progress of their work
to see that safety and health procedures are followed. A safety meeting is
held of all workers twice each year and other special meetings may be called if
warranted by a particular problem encountered. Workers receive medical examinations at least once a year and worker exposure to chemical substances is monitored
through the use of personal sampling devices.

The BASF interviews in Mexico confirmed that the safety and health policies
and procedures followed were the same as those used in the Federal Republic of
Germany. No instances had been found where the standards required by legislation in Mexico were more stringent than those of BASF, while the converse was
frequently found to be true. Mexican workers were given annual medical examinations, as were those at headquarters' operations. Mexican workers received
initial safety and health training as well as training each month dealing with
particular safety and health problems, including how to respond to emergencies
and cardiopulmonary resuscitation instruction. Mexican workers who were a part
of the plant fire brigade received training each week in fire protection as well
as emergency rescue and treatment procedures.

The occurrence of a safety and health problem thought to be significant in
any operating plant throughout the world must be recorded and reported weekly to
BASF headquarters. The headquarters staff analyses the problem, makes suggestions if necessary for its solution and communicates this information to all of
the operations throughout the world. A telex system is operated 24 hours a day
to answer queries from subsidiaries regarding safety and health problems.
When workers must be treated outside a BASF plant for a job-related injury or
illness, full written information regarding the problem is prepared by BASF
physicians and sent with the worker to the hospital. Such information is often
vital for ensuring proper treatment of the worker at the hospital.

Merck and Company, Inc.

The occupational safety and health programme of Merck and Company is headed by a Corporate Director of Safety and Industrial Hygiene. The corporate staff develops policies and procedures and monitors the performance of the international and domestic divisions. It is the policy of Merck and Company to establish control limits for safety and health hazards. The controls meet or exceed the national requirements in each of the countries in which they operate. In the event that the Merck and Company safety or health standards exceed those of the national authorities, then the more stringent standard of the enterprise is followed.

In addition to the extensive written procedures defining the safety and health programme of the enterprise, a unique computer system is used to store information regarding the formulations of all of the products manufactured by the enterprise. This provides readily available information about any possible health hazards to the workers. The safeguards which are to be followed in handling the materials are also stored in the computer and are included in any printout of operating instructions for formulating the product. Thus, each worker must view the safety and health precautions in handling the product in order to perform the formulation operation. The system is used in both the domestic and the international pharmaceutical divisions.

In an effort to promote a wide exchange of occupational safety and health information among the various operating units of the enterprise, two types of meetings are held for the safety and health staff. The corporate staff conducts a major seminar in the United States for the exchange of safety and health information with some 13 different locations represented each year. The International Division holds frequent regional seminars in Europe, the Far East and Latin America to exchange safety and health information.

All safety and health problems considered to be significant are reported to the safety and health staff of the international division, which in turn reports the information to the corporate staff. As an example of this type of information exchange, it was noted that an explosion had occurred in a piece of equipment involved in the drying of powders which is called a fluid bed dryer. As a result of this experience, an analysis was made of the source of the problem and information was rapidly transmitted to all of the subsidiaries where such equipment was utilised. In order to minimise problems of this nature, all of the equipment used in manufacturing operations is specified by the International Division staff for use in the Merck subsidiaries.

Because all of the products produced are pharmacologically active materials, reliance is placed on the investigations conducted by the research scientist responsible for developing the product regarding any potential safety and health problems in the manufacture of the product. The research scientist's evaluation is supplemented by a full review in the operating division of the safety and health measures necessary to manufacture the product, relying on corporate guidance as needed. In order for the pharmaceutical product to be licensed for sale in the United States as well as most other countries of the world, an extensive programme of testing must be conducted. The corporate safety and health staff is then able to utilise this information in establishing allowable exposure limits for workers who are involved in producing the product.

One of the cornerstones of the safety and health programme in the United States operations is the preparation of a job safety and health analysis for each job conducted within the enterprise. Using the occupational safety and health

information supplied by the corporate staff, the production supervisors in each
area, in conjunction with the industrial hygiene and safety staff, prepare the
job safety and health analysis in a written form which then becomes the operating
instructions for workers.   The supervisor must perform an annual review of the
written job safety and health analysis and update as necessary.   The medical
staff in each facility establishes a health surveillance programme with bio-
logical tests and medical examinations specifically designed for the potential
hazard and consistent with the degree of hazard to which the employee may be
exposed.   All employees are examined once each year, and those employees working with viral or bacteriological materials in laboratories may be examined as
frequently as once each month.

A safety and health committee comprised of workers and management representatives did not exist in the facility visited in the United States.   However, the
employee union within the facility regularly consults with the management regarding safety and health questions raised by workers.   In addition, the medical
staff informs workers about any hazards to which they may be exposed and offers
his or her full medical record and exposure information to any employee for
examination by their private physician.   It was found that one of the subsidiaries makes use of an independent joint occupational health service and that
this system is preferred by the workers in particular, as it eliminates any
suggestion of partisanship.

Regular safety and health audits are performed in all of the operating units
of the international division as well as the operating units in the pharmaceuticals
division within the United States.   Safety and health staff members from one
facility are selected to travel to another facility to perform the audit.   They
are accompanied by representatives of the workers as well as the management of
the facility during the audit.   It is believed that this system serves a very
useful purpose in encouraging compliance with the safety and health practices
of the enterprise as well as promoting a free exchange of safety and health
information among the professionals of the multinational enterprise.

Royal Dutch/Shell Group

In describing their safety and health activities, the representatives of the
Shell Companies noted that the safeguarding of the working environment was linked
with that of the environment as a whole.   Their concern went beyond compliance
with the requirements of the relevant legislation and, through their corporate
policies, took the form of an indirect control over the safety and health activities of all of its subsidiary companies.

Through their service companies, Shell has prepared an overall policy statement on safety and health, which is transmitted to all the operating companies.
Some of these operating companies are largely self-sufficient in information
regarding occupational safety and health, and they develop their own safety
policies, which have to satisfy the overall requirements of the Group policy
statement.   Other operating companies may need a substantial amount of advice
and assistance in order to implement the policy statement, and this is freely
available to them from the service company concerned.

The safety and health policy approach of the Shell Group was found to be
followed by the Shell Oil Company in the United States in dealing with various
operations in the country.   The Shell Oil Company corporate staff acted to produce policy and provide advice and assistance to the exploration, production.
refining and petrochemical components of the Shell Group.

The decentralised approach to safety and health within the Shell Group was observed in a visit to the Wood River, Illinois refinery, where the safety and health staff stated that each individual plant develops its own safety and health procedures under the general guidance of the service company. Each employee receives training in safety and health upon employment and is given an employee handbook which includes the major safety and health regulations which must be followed. In addition to the general procedures, each individual operating unit within the refinery has specific safety orders issued by management which must be followed by the employees operating the unit. The employees in each operating unit meet once each month to discuss safety and health matters. A safety and health committee consisting of seven workers and five supervisory members meets once each month to review any important safety and health matters as well as act as a forum for workers to present any questions which they have regarding safety and health procedures within the facility. The safety and health committee is given three days of special training each year in addition to the regular meetings. Although the individual safety and health procedures are developed within the individual facility, a uniform requirement exists to report accident investigations and other safety and health information to the corporate level.

The Pernis works of Shell Nederland Raffinaderij BV, the Shell operating company in the Netherlands, was found to operate with the same decentralised approach to safety and health as was found in the Shell Oil Company in the United States. Safety and health procedures were developed for the facility using the overall guidance from the Shell service company (Shell Internationale Petroleum Maatschappij BV). The safety and health committee was constituted according to the legal requirements, having two workers who were members of the works council, two members from the management of the facility, the medical officer, the safety officer, and three additional workers designated by the works council as well as three additional management members designated by the management of the facility. The committee serves as a forum for discussing questions regarding safety and health and health, and will take on additional responsibility with the implementation of the new government statute of 1983. It is noteworthy that the committee had already functioned for several years, even though not required by the previous statute.

Training of workers to follow proper safety and health procedures was considered a very important activity. Employees that were selected as supervisors spent several months as practical instructors, teaching the employees within the production area safety procedures and monitoring the employees to ensure that the procedures were being followed. Of course, this programme also resulted in the new supervisor having a thorough appreciation of safety and health matters.

All employees receive an annual medical examination, and those working in areas where contact can occur with chemicals are monitored with examinations as frequently as once every three months. The practice is followed of assigning a particular medical doctor to monitor a group of workers over a period of time, rather than having all of the medical staff share the responsibilities jointly. It was believed that this practice would provide for closer monitoring of the employees, with the likelihood that important changes in any health parameters would be more easily detected.

The safety and health programme within the Shell Petroleum Development Company of Nigeria, Ltd. was found to be well organised and effective, again following the decentralised approach noted earlier. However, in this instance, more direct reliance was placed on the safety and health practices manual developed by the Shell service company than was found to be the case for the

operations in the United States or the Netherlands.   The more extensive use of
this Shell manual is understandable, given that the operations are in a developing country as well as the fact that the operations are widely spread over a
wide geographical area.   The safety staff was well organised with a co-ordinator
in the Lagos office and operational safety officers distributed throughout the
various operating branches.   One section of the Lagos office was devoted to
carrying out safety audits throughout the country.

A fully equipped medical clinic with a number of physicians operating under
a chief medical officer was available for the routine medical examinations of
workers as well as treatment of injuries or illnesses which may occur to workers.
Provision was made for a special air evacuation of workers to the clinic in cases
of emergency.   A smaller secondary clinic also operated in another part of the
country in order to provide better access for the workers to medical facilities.

The safety staff regarded the Nigerian safety and health requirements as
minimum standards and noted that the Shell standards they followed went well
beyond these local requirements.   Regular training in safety and health occurs
at the levels of department heads, supervisors and workers.   Nigerians are
trained in the US for special safety responsibilities such as fire protection and
traffic safety.   Although no formal safety and health committees consisting of
workers and management exist, the safety department endeavours to maintain an on-
going dialogue with workers regarding safety and health matters.   This dialogue
is promoted through contests and a suggestion system.

It was stressed that safety performance is evaluated as any other management
responsibility and that safety performance reports were made to the parent company
in The Hague.   Special reports are made for serious injuries or fatalities in
addition to the periodic reports.

The safety and health performance of local contractors and subcontractors
performing work for Shell in Nigeria was a concern of management who stated that
very few of these contractors have developed a safety and health programme and,
in some cases, safety awareness is non-existent.   This is a good illustration of
the need for safety training and propaganda in the developing countries and is a
problem which requires the Shell safety officers to work with the contractors to
work with the contractors to help improve their performance.   One of the legal
requirements of the contracts signed by Shell-PD with its contractors is compliance with the Shell safety and health practices.

As noted in the safety and health reports from the other companies studied,
many accidents are linked with maintenance operations.   This can perhaps be
explained by the fact that these workers frequently operate alone or in small
groups and have to display a greater degree of judgement in adapting to a variety
of hazards than is required in the case of the more routine nature of the hazards
encountered by regular production workers.

The Bechtel Companies

The safety and health programme of the Bechtel Companies can be characterised
as a centralised programme.   The central safety and health staff have prepared
two manuals which serve as detailed guide-lines for conducting their safety and
health programmes.   One of them is concerned with operations within the United
States, having specific coverage of the way in which the Bechtel programme should
relate to the national authorities (OSHA).   A second manual has also been

prepared for international operations which contains the same material except for the deletion of material specific to compliance with the requirements of OSHA in the United States.

There is a clear chain of command within the central safety and health organisation with two major heads, one dealing with domestic operations and the other dealing with international operations. In practice, the site safety personnel are managed by a site safety manager who reports to the overall site manager for the construction project. However, the recruitement, training and assignment of safety and health personnel is controlled by the central safety manager. The responsibility for safety and health within a given area of a construction project clearly rests with the line superintendent or foreman for the area. The safety and health specialists are available to provide advice and to monitor the performance of the line supervisors in attaining a safe and healthy workplace.

The priority assigned by the site manager to safety in all of the construction operations was well illustrated by his description of the staff review procedure used at a major construction site with approximately 6,000 employees in the United States. He noted that all construction sites began with a build-up of activity which peaks at some point and then the employment declines until the project is fully completed. He explained that his most important criterion for selecting the workers as well as foreman and superintendents to be discharged as the employment decreased at the site was their safety and health performance, i.e. those with the worst safety records were the first to be discharged. Using such criteria obviously creates a powerful incentive for the foreman and supervisors at a site to train their workers carefully and to insist that safe practices be followed. Bechtel typically contracts with a local group of physicians to provide the necessary medical services at a given work site. These physicians work under an overall policy established by Bechtel's central medical staff, taking into account the local practices and procedures followed in each country.

In addition to the detailed safety manuals which are to be followed, Bechtel insists that every construction project have a Safety Action Plan which must be completed prior to the initiation of the project. This plant then serves as the overall guide for the operation of the safety and health programme at the construction site. The initial training of workers varies at each construction site, depending on their previous experience within the construction industry and the requirements of the national authorities. Tool-box safety meetings are held weekly with all workers on the construction site throughout the life of the project. These meetings are conducted by the foreman or superintendents with support as necessary from the responsible safety representative at the site. The site manager in conjunction with the safety manager conducts weekly meetings with all of the superintendents in order to discuss any special hazards which will be encountered during the next phase of the construction project.

Specific goals are established each year within the different Bechtel Companies for safety and health performance with the result that the injury frequency as steadily declined during the period of 1963 through 1982.

Bechtel echoed the concern of many companies regarding the supervision of subcontractors at the work site. Bechtel follows the same procedure as the Shell Group, stipulating in the contracts signed that safety and health performance acceptable to Bechtel must be maintained or the contract will be terminated. It was noted by the management at the construction site visited in the United States that a particular subcontractor had been suspended after a fatal accident and was not reinstated until the subcontractor's site manager had been replaced by a person having a much higher level of experience and motivation in the area of safety and health.

A visit to a Bechtel construction site in the United Kingdom confirmed that the safety and health procedures followed were identical with those in the home country. Worker representatives stressed the value of the weekly meetings held on the site to discuss safety matters. It was also noted that relations with the inspectorate were good but that inspectors seldom found it necessary to visit the work sites.

BICC Group

It has been noted earlier that the United Kingdom Health and Safety at Work Act requires that each enterprise have a clear statement of safety and health policy and that the means for carrying out the policy be fully elaborated. The health and safety manual of BICC begins with a statement of the health and safety policy for the group, and continues with a clear delineation of the respective responsibilities for carrying out the policy. It is therefore a good example of the type of compliance which the Health and Safety Executive expects in response to the policy requirements of the Act. Implementation of the BICC health and safety policy begins with an individual member of the Board being assigned the responsibility for health and safety policy and its implementation. The group medical officer has been assigned the responsibility for safety and health by this member of the Board, who in turn is in charge of three senior safety officers from different plants who are designated as group safety co-ordinators. The duties of the line and staff personnel of each group and units within each group are then clearly stated in the health and safety manual. In general, the responsibility for ensuring that the safety and health of workers is safeguarded rests with the line managers, who in turn receive the necessary staff support from specialists in health and safety. The health and safety responsibilities of the workers are also clearly stated in the policy. The three senior safety officers perform safety investigations and audits in the different plants to ensure that the policy is implemented.

The health and safety policy of BICC requires that joint management-worker health and safety committees be established. In general, these committees are established on a departmental level in the larger factories, but may be established for an entire factory in the case of the small factories within each Group. Worker members are nominated from each union within the department or factory to serve with the medical and safety representatives as well as members of management. A Group safety and health committee meets at least once every two months to consider a variety of topics within the overall realm of health and safety within th enteprises, producing written minutes of the meeting which are distributed to all of the members of the committee. Special working parties are set-up to deal with specified problems where an investigation is performed and a report made back to the full committee.

Workers receive thorough initial and follow-up medical examinations, the frequency of which varies in accordance with the type of hazard to which the workers may be exposed.

In general, the safety and health programme seemed to work well, with the one problem mentioned being the necessity to maintain close co-ordination between the various plants and groups in order that the policy be interpreted consistently. The Balfour Beatty, Ltd. subsidiary in Nigeria operates with the overall safety policy followed throughout the Group, supplemented by "Speciality Regulations" prepared as directives to fit the local conditions in Nigeria. As an example, the problems which have been described earlier regarding traffic accidents required special directives to deal with these within the company operations.

Because Balfour Beatty is acting as the overall project management for
three other overseas construction companies, the principal activities are concerned with monitoring the safety and health programmes of the contractors to
ensure that they comply with the contractual requirements. Problems exist in
Nigeria as in most Western countries with the legal implications involved in
directly supervising a subcontractor, i.e. the general contractor (Balfour Beatty)
can become legally liable for damages if accidents occur when direct supervision
is exercised over the employees of subcontractors. Therefore, it is necessary
for the overall contractor-manager to monitor the safety and health performance
of the subcontractors and in turn to criticise this performance as necessary,
but to avoid having a direct hand in the supervision of the workers of the subcontractor. The widely dispersed work sites involved in this overall project
requires close informal co-operation among the safety specialists and
engineers of Balfour Beatty and the three major subcontractors. The Medical
Officer/Safety Specialist is in radio contact with work sites to provide any
special guidance or action requested.

Special self-contained clinics are moved to work sites to provide medical
care. Medical evacuation flights are always available for seriously injured
or sick workers.

Volkswagenwerk AG

As might be expected because of its technical sophistication and its position
as one of the large employers in the Federal Republic of Germany, Volkswagen
places a high priority on the development of an effective safety and health programme throughout the company. The annual report of Volkswagen for 1981 makes
a specific reference to its goals for continuing to improve the safety and health
performance of the overall enterprise.

Volkswagen follows the requirements of the Occupational Safety Act in the
Federal Republic of Germany which was described earlier. It also endeavours to
introduce changes designed to add to worker satisfaction as part of the humanisa-
tion of work in line with the spirit of the Basic Law (Constitution) of the
Federal Republic of Germany. Safety stewards and occupational safety and health
committees are appointed in every factory. The operation of these committees,
in conjunction with the operation of the works council (Betriebsrat), provide
the workers with ample opportunity to participate in reviewing the occupational
safety and health conditions which exist in their workplaces, and in turn to work
with the employer to improve these conditions. Of course, the combination of
standards produced by the national authorities as well as the Mutual Accident
Assurance Associations (Berufsgenossenschaften) provide valuable guidance for
both the management and workers of Volkswagen in fulfilling the commitment of
the company to the maintenance of a high occupational safety and health record.
The appointment of trained safety engineers as well as medical officers in
accordance with the statutes provides both the management and workers with the
necessary technical information, as well as the required leadership to achieve
the established safety and health goals. Finally, the dual labour inspectorate
functions performed by both the constituent states (Länder) of the Federal
Republic and the Berufsgenossenschaften provide a continuing audit of the degree
of compliance achieved with the full range of statutes, ordinances, regulations
and other standards.

Volkswagen de Mexico is not considered to be under the direct managerial control of Volkswagen headquarters in Wolfsburg, even though the subsidiary is wholly
owned. The technical distinction was made that Volkswagen-Wolfsburg is not a

holding company, but nevertheless a major technical input is made to the operations of the various subsidiaries.   Because the technical managers in Volkswagen de Mexico had received their training at the Volkswagen plants in the Federal Republic of Germany and usually returned there after a few years of service, it was clear that they were very heavily influenced to follow the same standards in the plant in Mexico.   In all of the instances observed, the standards followed by Volkswagen in Mexico exceeded the requirements of the national standards.

In order to obtain the skilled workforce required for a major operation in Mexico, it was necessary for Volkswagen to undertake extensive training of local workers.   These workers are typically recruited at approximately age 16 and receive three-and-a-half years of technical training, after which they may decide whether to become employed by Volkswagen or to seek other employment.   Approximately 70 per cent of the trainees elect to accept employment with Volkswagen. Of course, the major training given the workers provides an excellent opportunity to also train the workers in safe practices while they are receiving their technical apprenticeship.   Workers are required to demonstrate a knowledge of the safety and health rules through passing an examination, after which they each sign a card pledging themselves to follow the safety and health rules of their employer.   However, despite the special training given the workers, it was noted by management that certain cultural differences existed between German and Mexican workers regarding attitudes towards risks in the course of their employment.   This was attributed to overall cultural views regarding life and death and the working of "fate".   It was noted that the workers' religion played a major part in their attitude toward occupational safety and health.   Each major work area in the Mexican plant contained an altar to the Virgin Mary, before which workers regularly asked for protection.

As noted earlier in the discussion of the occupational safety and health statutes of Mexico, the requirement to have an occupational safety and health committee exists within each establishment.   Workers have an opportunity to bring any questions regarding safety and health to the committee and receive an answer.   The Committee meets once each month to discuss safety and health issues and written minutes of the meeting are produced.   These minutes were said to be reviewed by the federal labour inspectors during their periodic visits to the facility.   In addition, visits were received from the staff of the Social Security Institute regarding occupational safety and health questions.   Visits within the facilities confirmed that the occupational safety and health standards followed were in all cases equivalent to those followed within the Federal Republic of Germany.

The Volkswagen subsidiary in Nigeria performs assembly operations only, with many of the components being imported, and the remainder being obtained from Nigerian subcontractors.   In general, it was found that the Nigerian national standards regarding occupational safety and health were not as strict or as detailed as those of the Federal Republic of Germany.   The Wolfsberg factory rules were being translated into English in toto for use in the local plant. The same situation was found to exist in Nigeria as was found in Mexico regarding the indirect application of safety and health standards through the exchange of personnel that had been trained in the Federal Republic of Germany.   Thus, even though the headquarters operations did not provide specific directives regarding the occupational safety and health standards to be followed in Nigeria, in reality the occupational safety and health standards followed were very similar because of the technical training of the personnel at home regarding safety and health matters.   A medical doctor and safety engineers provide the necessary advice and services for the operation of the safety and health programme.

All employees are trained regarding occupational safety and health upon assignment and obtain an employee handbook containing the occupational safety and health rules to be followed and for which they sign. Employees may receive up to three years of vocational training prior to their employment, including safety and health as was the case in Mexico. Selected employees may also be sent abroad to other Volkswagen locations for additional training, providing further opportunities to observe the functioning of an established occupational safety and health programme. Still, during a visit to the factory it was noticed that difficulty is experienced in motivating workers to wear respiratory protection in the joint spray booths.

Three-day seminars are conducted periodically for supervisors and managers to assure that they understand the safety and health practices and procedures to be followed. A safety committee consisting of shop-floor representatives from each department of the factory as well as the safety engineer meets periodically to review safety and health matters, and when necessary bring them to the attention of the managers involved. Members of the safety committee receive a special one-week safety course in conjunction with a government training programme in which the ILO is presently involved in Nigeria. Members of the safety committee are also given training in their duties by Volkswagen.

Xerox Corporation

Xerox has a central corporate staff for occupational safety and health as well as environmental safety and health matters. The corporate staff develops a uniform policy for occupational safety and health which is followed uniformly throughout all the Xerox subsidiaries. The policy is clearly stated and detailed procedures are established for the control of safety and health hazards associated with the products Xerox manufactures, the work environment within the plant and for the general environment exterior to the plant. The Xerox corporate standards are compared throughout the world with the standards produced by the national authorities and the more stringent requirements of the two are followed. In general, Xerox standards were found to be more stringent than those of the national authorities in any of the countries in which Xerox subsidiaries operate. Instances were noted where Xerox has established standards for the control of chemical hazards for which no standards have been established by the national authorities in the United States or other industrialised countries, as well as some standards where exposure levels were set well below the levels permitted by the national authorities in the United States or other countries. They have established what they call an "Action Limit", which is 50 per cent of the accepted MAC value, and if this limit is reached in any operation, preventive action must be taken.

Expertise exists within the Xerox corporate staff in the areas of occupational safety, occupational medicine and industrial hygiene. Xerox also has arrangements with some nationally known toxicologists to supervise the toxicological studies presently being conducted to test materials associated with Xerox products or Xerox manufacturing systems. In addition, Xerox has made a major investment in a computerised system for maintaining records of employee injuries and illnesses as well as exposure to any chemicals found in their work environment. It will be possible with this system to conduct toxicological and epidemiological studies for Xerox employees on a national or world-wide basis.

Each operating division of Xerox is expected to develop the appropriate procedures to implement the corporate policy. The line managers in each division are responsible for establishing the necessary procedures to implement the policy, but they are ably assisted by safety engineers, fire protection engineers, industrial hygienists, occupational physicians and occupational nurses who provide leadership, advice and audits of the performance of individual unit managers. All new or revised plant processes and equipment must receive a review and approval from the responsible safety engineer, fire protection engineer, and/or industrial hygienist before they are permitted to be installed within a plant.

Within the United States operations visited, management and union officials had voluntarily formed an occupational safety and health committee consisting of five members of the workers' union and five members of the plant management. It was found that the committee functioned as an effective forum for the discussion of occupational safety and health problems and as a means for workers to obtain information regarding any questions they had regarding occupational safety and health matters. The collective bargaining agreement was subsequently revised to incorporate provisions related to the operation of the committee.

Workers have full access to any records compiled relating to their medical condition as well as their exposure to any chemical substances found in their work environment. Medical and environmental monitoring is performed for all potentially exposed workers, with the interval for monitoring determined by the relative degree of exposure of the workers and toxicity of material. One of the workers interviewed noted that he was being requested to provide a urine sample for analysis once each month and that he was regularly told of the results of these tests.

Xerox also operates a major programme to encourage workers to observe good safety and health practices when they are away from their workplace. Employees are encouraged to equip their homes with fire extinguishers and smoke alarms, where the cost of these items is partially supported by their employer. In addition, a major facility exists where employees can be examined with regard to their level of physical fitness and a programme is prescribed for using the equipment within the facility to encourage employees to maintain an acceptable level of physical fitness.

The same problem of employee safety during operation of vehicles on public roads was observed in Xerox as with some of the other multinationals interviewed. It was noted that Xerox employed more than 13,000 technicians within the North American operations of the corporation to service machines. Special safety awareness and awards programmes had been conducted for these 13,000 service technicians, leading to a reduction in the incidence of accidents from 19 per million miles driven to 11. All hotels in which Xerox employees stay or conduct meetings must have approved fire safety systems.

Rank Xerox is the operating subsidiary of the corporation in the United Kingdom. The operations of this subsidiary follow the requirements of the Health and Safety at Work Act regarding the appointment of safety and health committees. As noted earlier, Xerox fulfils the requirements of the Act for a corporate policy and programme regarding occupational safety and health matters. Responsibility for implementing the corporate safety and health policy through the development of detailed procedures rest with the line management as in the headquarters operations. Occupational doctors and nurses as well as safety engineers are employed to provide specialised advice and services as well as to monitor the overall performance of the programme. Data must be regularly transmitted to the headquarters operations regarding serious accidents as well as the safety performance over the preceding three months' period. The data is analysed and corrective action required as necessary.

In addition to the worker representatives on the plant safety committee, a system exists by which a floor superintendent or equivalent, known as a "safety landlord", is detailed to be responsible for safety in a specific area or the Welwyn Garden City plant.   The system provides a rapid means for workers to bring any safety concerns they may have to the attention of management and the safety and health specialists.   These "safety landlords" are also required to transmit information to the workers.   The workers interviewed considered this system helped to answer their questions regarding occupational safety and health as well as enabling them to draw attention to any changes they considered necessary.

The Xerox operations as observed within Mexico were concerned with the refurbishing of used Xerox units.   The assembly of new units had recently been transferred to a new facility some distance away which was beginning operations near the time when the interviews were being conducted.   Some unique hazards were associated with the refurbishing of used units as compared to the assembly of new units.   One hazard concerned the exposure of workers who were involved in cleaning the accumulation of toner from the inner workings of the machines prior to the refurbishing operation.   The cleaning operation is usually accomplished by spraying a solvent under pressure throughout the machine in a specially constructed booth with exhaust ventilation.   The solvent mist emitted during the spraying operation as well as any toner or other residues which may become airborne as a result of the pressurised spraying constitute a health hazard if breathed.   Another environmental hazard exists if the electrical components which contain polychlorinated biphenyls (PCB) are not properly disposed of when removed, thus requiring special arrangements to be made.

In accordance with Mexican law, a safety and health committee consisting of workers and management operated in the plant to provide a forum for the discussion of occupational safety and health problems.   Monthly meetings of the safety and health committee were held and the written minutes of the meetings distributed in accordance with the requirements of the labour inspectorate.

The management of Xerox de Mexico noted that considerable difficulties existed in hiring competent safety engineers.   Recruitment was in progress at the time of the interview to fill such a position and it was hoped that someone might be found in the near future.   (It was remarked that another MNE operating in Mexico, the General Motors manufacturing facility, had been attempting for several months to hire a qualified safety engineer.)   A safety committee whose members were trained to administer first aid existed within the plant visited where machines were rebuilt.   An occupational physician and nurse visited the plant from time to time to perform various types of medical screening.   Any worker seriously injured within the plant requiring immediate medical attention is transported to a clinic of the Institute of Social Security which is located very near the plant to serve the industries in the area.

Workers receive training in occupational safety and health, with emphasis on the use of personal protective equipment and proper techniques for manual handling of materials.

Lack of space within the plant added to the occupational safety and health problems.   Xerox noted that they had petitioned the government authorities for permission to expand their facilities but had so far been unsuccessful because of the government policy to relocate industry further away from the Federal District of Mexico.   Overcrowding of the facility made it difficult to keep aisles free of obstructions and led to some work where solvents were used to be accomplished without adequate ventilation.   (It was learned at the time this report was going to press that the Mexican Government had granted the permission requested to expand the facility.)

The large number of service technicians driving cars or riding motorcycles on the city streets in the course of their work was considered to be one of the most serious continuing safety problems. Service technicians receive special training in defensive driving techniques as well as having an incentive scheme to motivate safe driving behaviour.

Brown Boveri and Company, Ltd.

The central BBC management in Baden, Switzerland, stressed that the overseas subsidiaries were completely independent with regard to the implementation of safety and health standards and practices, taking no direction from the Swiss parent company in this respect. If a subsidiary company requests information regarding occupational safety and health, it would be supplied, but no require-ment exists for the subsidiary to follow any advice which is requested from the parent company. Fire prevention information is exchanged on a regular basis between the parent company and subsidiaries. It was stressed that the parent company exercises no control over implementation of occupational safety and health measures within the various subsidiaries.

With respect to the operations in Switzerland, all safety and health activi-ties within the company are dealt with through the safety committee which includes workers from each department in the plant, a representative from the safety department, a management representative and specialist advisory members covering occupational medicine, electrotechnics, environmental protection, fire protection, the purchasing department, chemicals and toxic substances and radiation protec-tion.

Safety and health standards at the Baden works were described by management as being above average for Switzerland. This opinion was also held by the officers of the SUVA who were interviewed.

The BBC subsidiary in Mannheim, in the Federal Republic of Germany, was found to closely follow the model of occupational safety and health which has pre-viously been described for the BASF and Volkswagen operations. Since no direc-tion is provided by the parent company with regard to occupational safety and health, the works safety organisation issues directives regarding compliance with the various laws, ordinances, codes and standards. In addition, special standards or data sheets may also be issued for particular problems.

Because of the large number of workers employed in Mannheim (36,000) it is necessary to have an extensive safety service. The safety service includes 13 graduate safety engineers who are supported by 70 safety specialists. It is the responsibility of these safety engineers to ensure that appropriate standards are developed and followed as well as to monitor safety performance within the estab-lishment. Plant inspections with worker participation are regularly conducted. Approximately one worker out of every 20 receives special instruction in occupa-tional safety.

The works council has the important responsibility of reviewing the equipment and processes within the factory to ensure that workers' safety and health needs are adequately catered for by the management of the enterprise. Agreements are made between the works council and management and between the unions and manage-ment which may supplement the rules and regulations of the state and industrial insurance associations.   Workers receive medical examinations and treatment in accordance with the statutes described earlier for the Federal Republic of Germany.

Brown Boveri Mexicana, SA is a relatively small subsidiary consisting of approximately 100 factory workers and 100 support and sales staff persons.   The management of the enterprise confirmed that no safety and health recommendations had been received from the parent company in Switzerland.   The safety and health policy was stated as being based on compliance with the Mexican safety and health laws, as well as with some individual standards for the plant which had been developed as a result of experience gained by the technical staff in Switzerland.

In accordance with Mexican law, a safety and health committee was organised for the factory and its monthly meetings provided an opportunity for the workers to bring their concerns regarding occupational safety and health to the attention of the management.   As an example, the management representative being inter-viewed noted that the workers had requested in a recent meeting that no painting be done within the factory except within the ventilated paint booth.   Management considered that it was too difficult to comply with the request.

Because of the small size of the plant, no safety or medical specialists were appointed.   However, workers as well as members of the management were trained in administering first-aid procedures, and more extensive medical help was avail-able at a nearby clinic of the Institute for Social Security.   Every worker undergoes a complete medical examination twice a year in conjunction with the Institute for Social Security.

Measures of safety and health performance
for MNEs participating in the study

Ideally, it would be desirable if a uniform measure of the safety and health performance of the multinationals participating in the study could be devised which would permit precise comparisons to be made between the different countries in which they operate as well as with other enteprises having similar character-istics.   Unfortunately, a wide variety of practices exist for recording and reporting data to indicate the safety and health performance of an enterprise. The ILO and the International Social Security Association (ISSA) have attempted to provide leadership in establishing more uniform systems for this purpose. However, a great deal of diversity continues to exist and because of this and other methodological problems referred to earlier in this report, it was not possible in the present study to compare every participating multinational enter-prise on a common basis in terms of its safety and health performance.   The information which follows was obtained from a variety of sources, including statistical summaries in some instances, and general evaluations in others. Anecdotal information was used when no other sources were available.   The infor-mation is arranged for each MNE participating in the study, including statistical comparisons where these are possible between the MNEs in the study and between other similar enteprises within the same industrial sector.

In order to provide a background for interpreting some of the health and safety statistics, a brief description follows of some of the more commonly used measures for occupational safety and health performance.   The two most fre-quently measured parameters are the frequency and the severity of accidents. The frequency of accidents is usually related to a base of number of workers or number of work-hours.   A wide range of definitions exist for an accident:

(1) An unplanned event in the work process.

(2) An unplanned event resulting in property damage or an injury to a worker.

(3) An unplanned event resulting in an injury to a worker requiring more than simple first-aid treatment.

(4) An unplanned event resulting in an injury to a worker which requires his or her absence from the workplace for more than a given number of days, ranging from one to seven days, with three days' absence being the most common number of days used in the various countries. Weekend days are usually not counted unless the worker had been scheduled to work on these days.

Injuries are frequently referred to a base of 100 or 1,000 workers, or to a working period of 1 million work-hours. It is necessary to use some form of system for normalising the injury data, otherwise the enterprises with a large number of workers would appear to have poor records because of the largest number of employees exposed to the possiblity of having an accident.

It is more difficult to obtain a valid measure of the severity of an accident. If an employee is absent from his or her workplace for a given number of days as a result of a work injury from which they fully recover, then presumably a useful measure of the injury severity would be the number of days the worker was absent from the workplace. However, there is a wide difference of opinion among occupational physicians regarding the number of days that an employee should be permitted to be absent from the workplace in recovering from a given kind of injury. It was found in conducting this study that the occupational physicians in some countries were assigning a longer period of time for the recovery of an employee with an injury than would be assigned in another country for the same injury. Again, the rate of recovery from similar injuries may vary for different individuals. Such variations obviously distort the statistics used in attempting to measure the severity of injuries to employees.

Injuries which result in a permanent disability to a worker create an even more difficult problem in measuring severity. The amputation of an employee's finger may result in the same number of lost work-days as an employee recovering from an injury where no permanent disability occurred as a result of the injury. In order to have some form of accounting for permanent disabilities to workers, tables are developed which equate a given type of permanent disability to a number of lost work-days which are to be charged for the permanent disability. The problem is further complicated in the case of occupational diseases where a given percentage of disability is assigned as a result of the illness, raising the question of how many lost work-days should be assigned as a result of the occupational illness. Given an older worker and a younger worker with the same degree of disability resulting from an occupational illness, should more lost work-days be assigned as a result of illness to the young worker than the older worker?

Another problem related to the interpretation of health and safety statistics is the definition of a work injury. In many countries the employee has to have been injured on the premises of the enterprise, while other countries also include injuries sustained by workers while travelling to or from their workplace which are usually injuries sustained in traffic accidents.

The foregoing brief explanation of the statistical approaches to occupational safety and health has been included with a view to stressing the caution which should be exercised when examining the following data relating to the enterprises taking part in the study.

BASF

A very complete set of safety and health statistics for the Ludwigshafen works was provided, together with some statistics for other locations. Some of the highlights for the Ludwigshafen operations are as follows:

(1) Reportable work accidents (accidents resulting in death or three or more lost work-days) declined 24 per cent from 1980 to 1981.

(2) Total work accidents decreased by 14 per cent between 1980 and 1981.

(3) The number of reportable work accidents have decreased each year from 5,109 to 1972 to 1,100 in 1981.

(4) The hand was the most frequently injured part of the body, accounting for 42 per cent of reportable work injuries, followed by the foot with 18 per cent.

(5) For the first time in the five years of statistical data provided, no fatal work accidents were recorded for the entire year of 1981. The number of reportable work accidents per 1 million work-hours in the Federal Republic of Germany in 1981 was approximately 38 for all of industry, 27 for the chemical industry and 13 for BASF, Ludwigshafen.

Similarly, detailed statistics were not available for the BASF subsidiary in Mexico. The Santa Clara plant visited had experienced 11 reportable accidents during 1982 for a workforce of approximately 200 people. However, the record was much worse during the year 1981 when six workers died following an explosion in a chemical factory, together with five seriously injured and 12 slightly injured. Headquarters management noted that the generally good safety and health record of the location was marred by this major fire and explosion in 1981.

A general indication of the safety and health record of an enterprise in Mexico can be obtained from the rating assigned by the Social Security Institute on which the employer's payments to the system are based. Each enterprise is assigned a rating in terms of a high, middle or low level of safety and health claims experience within an industrial sector. The Santa Clara location had received notification immediately prior to the interview in early 1983 that their rating had been changed from the high category to the middle category. It would therefore appear that the occurrence of a major accident such as the one recorded in 1981 had resulted in a high level classification, while the adjustment in the rating in 1983 represented a recognition of the good record achieved during 1982.

To summarise, if the accident experience of the subsidiary is compared with that of the home operations on the basis of available statistics the subsidiary clearly has a worse record. However, the major accident which occurred in 1981 could no doubt be considered not being typical of the generally good record which had been achieved prior to that time. A reportable accident at the Santa Clara location was defined in the same manner as at the Ludwigshafen location, i.e. three days or more away from the workplace as a result of the injury. Using the statistics of number of reportable injuries per 1,000 workers, the Ludwigshafen location experienced a rate of 22 during 1981, while the equivalent rate during 1982 at the Santa Clara location was 55, which clearly shows the home operation in a more favourable light than the subsidiary.

Merck and Company, Inc.

Merck and Company is a member of a group of enterprises within the pharmaceutical industry which co-operate in the improvement of occupational safety and health. An index is computed for the overall group against which the individual members can then compare their safety and health records. The index of safety and health employed for comparison is a combined frequency and severity measure computed as follows:

$$\text{Frequency} = \frac{(\text{No. of reportable injuries}) (200,000)}{\text{Total number of work-hours}}$$

$$\text{Severity} = \frac{(\text{No. of lost work-days}) (200,000)}{\text{Total number of work-hours}}$$

$$\text{Performance index} = \sqrt{\frac{(\text{Frequency}) (\text{Severity})}{40}}$$

Referring to the discussion of measures of severity presented earlier, Merck and Company assigns a lost work-day value of 365 days when a fatality occurs.

Complete statistics were provided for all Merck and Company operations based on the performance index. Some of the highlights related in this study are shown in table 2.

Table 2: Merck and Company, Inc. - safety performance

Operating unit	Performance index			
	1980	1981	1982	1983 (first half)
Entire company	0.96	1.17	0.85	0.71
MS and D - Domestic	1.76	1.49	1.13	0.55
MS and D - International	0.50	0.63	0.49	0.47
Netherlands	1.12	0.38	0.45	0.81
Pharmaceutical related operations only				
MS and D - Domestic		1.10	0.78	0.41
Industry Group (USA)		0.87	0.76	0.52

Source: Data provided by Merck and Company, Inc.

The statistics for the company overall indicate an improving performance index since 1981. Performance in the Merck Sharp and Dohme domestic operations have been substantially worse than those in international operations, even though both divisions perform similar operations. Because of this imbalance, Merck management has worked to improve the safety of workers in domestic operations and the improvement is apparent for the first half of 1983. It is clear that progress is being made in the pharmaceutical and related operations within the United States as compared with the pharmaceutical safety group in the industry. Performance in the Netherlands was clearly superior to both the domestic and other international operations within the United States as compared with the pharmaceutical safety group in the industry. Performance in the Netherlands was clearly superior to both the domestic and other international operations until the first half of 1983.

Royal Dutch/Shell Group

According to headquarters management, health and safety statistics are provided annually from all operating companies to the service companies on a functional basis, i.e. exploration and production, manufacturing, etc. Performance indices are based on their calculations of frequency rates and severity rates. Since 1957, when figures were first recorded, the frequency rate declined from 28.4 to reach a plateau at about 14.4 over the years 1962 to 1966, then fell sharply to a new plateau of about 8.6 from 1972 to 1979. Since that time there has been a further downward trend to 6.6 in 1982. It was noted by the company that, while statistics can provide a general trend of safety performance, they place greater emphasis on detailed accident analysis reports and general safety audits, which enable deficiencies in safety procedures to be highlighted and provide lessons that can be communicated to all areas in order to achieve greater safety awareness.

In the case of the Pernis refining operation in the Netherlands and that of the exploration and production operations in Nigeria, figures were not quoted because of the difficulty of making meaningful comparisons between one type of operation and another. The factory inspectorate in the Netherlands considered that the safety and health performance of the Shell works was above average, and a similar view was expressed by the authorities in Nigeria.

Health and safety statistics available from the Shell Oil Company (US) indicated a total of 0.44 reportable injuries per 100 workers for 1981, and 0.32 for 1982, compared to an industry average of 1.39 for 1981 and 1.51 for 1982. A reportable injury in this case is considered as one where the worker cannot show up for work at his next regularly scheduled shift. No fatalities had been reported since 1976.

It is of interest, from the point of view of this study, to note the difference between the calculations of frequency and severity rates used by Merck and Company, Inc. (see above) and that of the Shell Group which was stated to be as follows:

$$\text{Total frequency rate} = \frac{\text{total number of disabling injuries} \times 10^6}{\text{total man-hours worked}}$$

$$\text{Total severity rate} = \frac{(\text{total of scheduled changes and actual time lost in days}) \times 10^6}{\text{total man-hours worked}}$$

Such differences clearly make meaningful comparisons difficult.

Data was not available regarding safety and health performance for the
Pernis works of the Shell operating company in the Netherlands.   However, the
factory inspectorate considered that the safety and health performance of the
Shell works in the Netherlands was above average as compared to the petroleum
and petrochemical industries.

Bechtel Companies

According to the information given, the administration of safety and health
within the Bechtel organisation is formed into seven divisions, five within the
United States and two covering overseas operations, each of which is in the
charge of a safety manager.   Statistics are drawn up on a group basis under the
responsibility of the group safety engineer.   Each division submits monthly
accident reports to this headquarters together with quarterly statistical returns
within a standardised framework.   Any serious or unusual accident would be
investigated by the divisional manager who reports back to headquarters.   A
study of the recent statistics shows that within the group there has been a clear
downward trend ever since the year 1968, and that reportable accidents have
decreased by an average over the period up until the present time of approximately
5.5 per cent each year.

The BICC Group

Statistics within the BICC Group include the recording of all minor injuries
and the recording of all serious accidents which are defined as those entailing
absence from work for more than three working days.   These statistics are kept
separately for each of four main companies, for the central department which
includes headquarters, and, within the four main companies, for each of some 50
plants or operating sections.   The widely ranging nature of the activities of
these plants or sections, which include heavy electrical equipment, electronics,
small components and cable manufacture, as well as the construction operations
of Balfour Beatty Ltd., makes comparison of one with the other invidious.   However, an across-the-board average of serious accidents indicates a decrease over
the years from 1980 to the present closely approaching 20 per cent.   In the case
of the Balfour Beatty construction operation in Nigeria (the subsidiary visited),
the ephemeral nature of this type of operation precludes any long-term collection
of statistics.   However, from information gathered, since the start of the
operation, there had been no fatal accidents and, apart from minor accidents
such as cuts and falls, the only incidents entailing absence from work of more
than three days had been a few cases of heat stroke or dehydration, some mild
attacks of malaria (in spite of prophylaxis) and two or three cases of dermatitis,
all of which had been successfully treated by the company doctor in its own
clinic.

Volkswagenwerk AG

Volkswagen-Wolfsburg provided a number of measures of occupational safety
and health performance which are as follows:

(1) The number of days lost as a result of reportable work accidents were
     293 per million work-hours.   (This rate is the second lowest among seven
     major automobile manufacturing enterprises in the Federal Republic of
     Germany.)

(2) The average number of days lost per accident were 12.0, which was again the second best record among the seven major automobile manufacturing enterprises within the Federal Republic of Germany.

(3) The number of accidents where the worker was absent for four or more calendar days as a result of the accident were 24.4 per million work-hours, again the second best record within the automobile manufacturing enterprises within the Federal Republic of Germany. A large volume of statistical information related to occupational accidents and illnesses was provided for operations in the Federal Republic of Germany. This type of information is very useful for developing preventive measures in a specific enterprise, but is too voluminous to be reproduced here.

The statistical information available from the Volkswagen operations within Mexico was reported differently than the information from Wolfsburg. Using an estimate of 2,080 hours worked per worker each year, the total of 1,193 accidents would amount to about 41 accidents per million work-hours. The rating system employed by the Social Security Institute to determine the employer's contributions to the system provides a general indication of the safety and health performance of an enteprise. Volkswagen had been classified for approximately six years within the middle level of the three levels of risk assigned within the industry.

In the case of the Nigerian operation of VW, the company kept its own accident records consisting at the present time of an accident register in which all cases, including minor accidents, were recorded at the factory clinic. Long-term statistics were not available because of the short period of operation, but a scheme had been drawn up and would shortly be put into operation. Up to the present time, there had been two serious (over three working days) traffic accidents and one due to a worker being trapped under heavy material while lifting. A number of minor accidents had been reported in the woodworking section of the plant. Occupational diseases were reported according to national requirements, but the country-wide collection of accident statistics is not yet well organised.

Xerox Corporation

Table 3 presents the safety and health statistics for Xerox. The statistics are all in terms of incidence rate per 100 workers for the various categories of data.

Table 3: Xerox safety performance, 1982

Safety statistic	Overall corporation	European operations	Mexican operations	US operations
Recordable cases	4.57	3.30	3.20	7.45
Days away from work cases	1.76	N/A	N/A	N/A

Source: Data supplied by company.

Table 3 shows that the subsidiary operations of Xerox had better safety records than the parent company in the US.

Brown Boveri and Company, Ltd.

The reportable work accidents reached a value of 49.4 per 1,000 workers during the year 1981 for all of the BBC operations within the Federal Republic of Germany. This value represented a 7.3 per cent decrease over the value of 53.3 reportable work accidents per 1,000 workers experienced during the year 1980. The number of reportable work accidents per 1 million work-hours during 1981 was 29.4, again representing a decline over the value of 31.5 reportable work accidents per 1 million work-hours achieved during 1980.

Statistics regarding occupational safety and health performance for the BBC operations in Switzerland were also obtained. These statistics are compiled on a different basis than in the Federal Republic of Germany, so any comparisons must be made with care. All injuries requiring medical care are reported, followed by a classification of these injuries into "normal" and "minor" categories. The rate of accidents requiring medical care during the year 1981 was 2.6 per 100,000 workers. Of this total, 1.4 accidents per 100,000 workers were classified as "normal", with the remainder (1.2) being classified as "minor" accidents.

No information regarding the safety and health performance of the BBC subsidiary in Mexico was obtained during the interview.

CHAPTER III

COMMUNICATION AND CO-OPERATION IN ACHIEVING
OCCUPATIONAL SAFETY AND HEALTH GOALS

The material in this chapter focuses on one of the principal concerns of
the study, the communication of information regarding occupational safety and
health between workers, employers and national authorities.   Each of the
sections which follow focus on one aspect of the overall information transfer
or exchange taking place.

Communication and co-operation between
headquarters and subsidiary operations

The information necessary to safeguard adequately the health and safety of
workers is growing each year.   This is especially true in the case of occupational illnesses where an increasing awareness of the diseases which can result
from exposure to chemicals in the workplace has made it necessary to invest large
sums of money to peform toxicological testing on a wide variety of products.
A comprehensive test of the toxicological effects of a given chemical substance
can cost US$200,000 or more.   If one considers the large number of products
which need to be tested, it is clear that the sharing of information regarding
the results of such tests is essential in order to co-ordinate the overall testing programme of a multinational enterprise and avoid unnecessary expense.
However, since this form of testing does require such a significant expenditure
of funds, the financial arrangements between headquarters and subsidiaries
operations regarding the value of the shared information must be seen within
the overall economic context of the company and the responsibility to the share-
holders in the respective countries.   This point was especially noted by the
Shell Group management.

Epidemiological studies to relate observed worker illnesses to the hazardous
material in the workplace represent another instance where the sharing of information between headquarters and subsidiary operations is frequently necessary.
In general, the larger the sample of workers which can be studied for exposure
to a workplace hazard, the more efficient the statistical analysis becomes.
This is especially true when one is attempting to develop a risk assessment for
a toxic material used within the company as a whole and it is necessary to estimate the incidence of the occupational disease for different exposure levels of
the toxic material in the workplace.   The ability of the headquarters operations
to compile medical records and exposure information for all of the subsidiaries
using a particular material would definitely enhance its ability to perform
epidemiological studies within the overall enterprise.   Xerox Corporation has,
for instance, ivnested a large sum of money in developing a computerised system
for handling the medical and exposure records for employees throughout the
Corporation.   Such large investments are necessary if a multinational enterprise is to have an efficient method of collecting medical and exposure information about employees for performing epidemiological studies.

Several of the mutlinational enterprises participating in the study (Xerox,
Bechtel, Merck, BASF and Shell) have an established system where an occupational
health or safety problem which is recognised any where within the overall enterprise structure is quickly notified to the headquarters unit for evaluation,
after which information regarding the problem and its solution is rapidly

communicated to all locations where the problem might exist throughout the enterprise. The operation of such an enterprise-wide system of collecting and exchanging information clearly benefits the enterprise in conserving both its human and monetary resources.

A number of the multinational enterprises participating in the study relied on a safety and health manual developed by the headquarters unit to communicate information about the manner in which the safety and health programme within the subsidiaries should be conducted. In some cases, the manual is supported by data sheets relating to dangerous substances used or to particular hazards. These again are developed by headquarters, regularly updated, and are available to all supervisors and workers. The use of these devices, which are relatively inexpensive, offers clear advantages for assuring a uniform and potentially effective approach to occupational safety and health throughout the enterprise.

Audit programmes were conducted by the headquarters unit of most MNEs to assure compliance with the safety and health manual throughout the enterprise. The safety and health staff of different subsidiaries are often exchanged in order to carry out these audits and to increase the exchange of information.

Some of the multinationals participating in the study used periodic regional meetings for management and for safety and health staff of the subsidiaries operating in one part of the world to exchange health and safety information, ranging in frequency from two to three times each year to meetings occurring once every two years. Such meetings also offer an opportunity for the head-quarters staff to correct any errors in understanding and implementing the safety and health policy of the enterprise as a whole, as well as an opportunity for the professionals present to share experiences, and discuss needed improvements in their respective programmes. Management motivation concerning safety and health is also improved through such exchanges because of the mutual reinforce-ment received from senior staff members. While it is desirable that the sub-sidiary units report their occupational accidents and illnesses to the head-quarters unit for analysis and comparison with other units, problems have been found to exist in establishing a uniform basis for the collection of this informa-tion because of the varied reporting requirements of the national authorities and the different conditions in the regions where the subsidiaries operate. Har-monisation of the requirements of the various national authorities where the collection of safety and health information is a goal of the ILO and ISSA. Progress attained in this respect will in turn make the exchange of such informa-tion between the various units of the multinational enterprise more beneficial. It would seem, however, that the enterprises themselves could usefully reinforce their efforts towards achieving greater harmonisation in the collection of statistical data.

Communication and co-operation between
MNEs and national authorities

In view of their international operations, involvement in new production processes and comparatively well-developed programmes, MNEs are in a favourable position to provide information to the national authorities in the countries where they operate regarding safety and health measures which need to be followed for the protection of workers. Especially those MNEs with centralised safety and health programmes have a broad base of experience on which to draw when preparing manuals or other forms of instruction for the guidance of the operating units of the subsidiaries throughout the world. As many MNEs operate with a well co-ordinated system of information, they are also well placed to

collect information regarding the laws and regulations governing labour inspection in all the countries where they are located.   This collection of information permits comparisons between the different regulatory approaches followed
by various national authorities and the result subsequently achieved.   In some
instances, the laws promulgated by the national authorities for occupational
safety and health are well written and complete, but the enforcement of these
laws by a well trained factory inspectorate is lacking.   In other instances,
the laws may be incomplete, even though the factory inspectorate has an adequate
enforcement ability.

Thus, many of the MNEs are in a position to share the following types of
information with the national authorities in the countries where they operate:

(1) Legal and regulatory structure for occupational safety and health in other
     countries.

(2) Experience regarding the value of a given approach to the establishment and
     enforcement of statutes for the protection of workers in a country.

(3) Scientific information available from their own experience regarding the
     relationship between a worker's injury or illness and certain conditions
     in the workplace.   (This information is particularly important in the case
     of chemical hazards.)

(4) Effective methods for training workers to observe proper occupational
     safety and health practices and obtaining worker participation in the occu-
     pational safety and health programme of the enterprise, based, again, on
     their own experience.

The MNEs are also in a position to, and in some cases do, assist in the
accident prevention activities of the national authorities and associations and
wish to be seen as setting an example as regards safety at work, traffic safety
and general safety awareness.

The following sections provide information obtaining during the study regarding the exchange of safety and health information between the participating MNEs
and the national authorities in the home and host countries visited.

The Federal Republic of Germany

When considering the occupational safety and health programme in the Federal
Republic of Germany, it is insufficient to take into account only the actions of
the governmental authorities as comprising the whole programme since the actions
of the Mutual Industrial Insurance Associations ("Berufsgenossenschaften") are so
closely interlinked and both play a major role in the overall occupational safety
and health programme within the country.   Therefore, the comments which follow
relate to the national authorities, the authorities of the constituent states and
the Mutual Industrial Insurance Associations.

It has been noted earlier that the accident prevention regulations of the
"Berufsgenossenschaften" form a major part of the requirements which employers
must follow to provide a safe and healthful workplace.   Through their contact
with the representatives of the Berufsgenossenschaften, multinational enterprises
have an opportunity of providing information about the need they see for regulations to protect the worker and the means for implementing these regulations.
Similarly, the MNEs, like other enterprises, also have the opportunity to express

their views to the federal and state authorities regarding the laws and ordinances promulgated for the protection of workers. The safety and health specialists of the MNEs can also discuss detailed measures for compliance with the safety and health laws, ordinances and regulations with either the inspectors from the constituent states or the "Berufsgenossenschaften". The MNEs, like other enterprises, are therefore in a position to provide information regarding the protection of workers, and in turn receive relevant information from their contacts with the national authorities and the "Berufsgenossenschaften".

Another major opportunity for information exchange with the authorities in the Federal Republic of Germany occurs when MNEs purchase equipment within the country or import equipment from other countries. The Guarding of Machinery Act establishes specific requirements for testing the safety of machinery prior to its installation in the workplace. The MNEs, like all other companies, either as manufacturers of equipment sold inside the country, or as purchasers of equipment either manufactured in the country or imported, must become aware of the regulations for inspection and testing of the equipment in order to have it certified as safe. This process allows for the exchange of views with the staff of the testing centres throughout the country regarding the technical means for compliance with the regulations. Information thus gained can then be transferred by the MNEs to the technical specifications to be met for equipment purchased in other countries where these do not have comparable requirements.

From observations made during the study, it does not appear that formal arrangements for the exchange of information between the authorities and MNEs or industry in general exist on a wide basis at least in the countries covered. Certainly in the case of new or proposed legislation, the custom of holding consultations with the social partners is common in most of the industrialised countries. In the United Kingdom, an advisory unit has been appointed which has no enforcement role but relies on voluntary co-operation with the enterprises to investigate and make recommendations on industrial accidents and diseases, which has been well-recieved by the employers, and which has reported considerable success as a result of its activities.

On the other hand, informal contacts were found to be common, usually at the level of the inspectorate. Inspectors, as part of their general task, both provide and receive technical information and transmit this where appropriate to other enterprises having similar difficulties or seeking advice. It was also found in some countries that technical staff of the government service participate in meetings of professional associations or similar bodies where these exist, and are called upon to assist in such activities as training seminars concerned with safety and health. In one case, it was noted that a special tripartite committee had been set up to study accidents in the construction industry.

The United States

The sources of regulations governing safety and health in the workplace within the United States are the Occupational Safety and Health Administration (OSHA) for all of industry except mining and milling operations, and the Mine Safety and Health Administration (MSHA) for these regulations. The procedures for producing regulations by either OSHA or MSHA are lengthy and involve several steps where information is specifically solicited from any employer affected by the regulations regarding the impact of the regulations in improving safety and health in the workplace as well as the cost of implementing the regulations. In principle, all major enterprises, including MNEs are specifically requested to comment on the need for a regulation to eliminate an occupational safety and health hazard as it is initially being drafted and again at the time that the complete regulation is proposed.   While the national authorities of the United States maintain extensive contacts with the national authorities of other industrialised countries in order to collect information about occupational safety and health hazards, it is recognised that the MNEs can also serve as an additional source of information regarding the occupational safety and health practices of other nations in eliminating workplace hazards.

The United States does not have a system comparable to the "Berufsgenossen-schaften" in the Federal Repbulic of Germany to propose regulations for the pro-tection of workers, but it does have an organisation which exerts a world-wide influence in the area of occupational health.   This is the American Conference of Governmental Industrial Hygienists (ACGIH), which has a membership limited to professional personnel in governmental agencies or educational institutes engaged in occupational safety and health programmes.   The Industrial Ventilation Com-mittee and the Threshold Limit Values Committee of the Conference are both recog-nised throughout the world for their expertise and contributions to occupational health.   The Conference publication entitled Threshold Limit Values for Chemical Substances in the Work Environment is widely used internationally and is often considered outside the United States to be an official publication, even though the Conference is not a government agency.   Research information from through-out the world is used by the Conference Committee to develop the Threshold Limit Values (TLVs).   Since the TLVs are used for evaluating worker health risk from chemical substances and physical agents in the United States as well as many countries of the world, multinational enterprises are frequently involved in exchanging information with members of the TLV Committee.

A variety of federal statutes require enterprises within the United States to supply information to the national authorities regarding the use of toxic materials in the workplace as well as those which are discharged into the air or water leaving the site of the factory.   Civil law also exists which requires disclosure of information about the hazards of almost any material used in the workplace.   A national system of Material Safety Data Sheets (MSDS) is volun-tarily used by manufacturers of these materials to inform the user of its chemical properties, hazards associated with the use of the materials under various conditions, and the precautions which should be taken to protect workers exposed to the material.

The Shell Oil Company (USA) and the Xerox Corporation have extensive pro-grammes of epidemiological and toxicological testing.   Both enterprises also have a policy of making public the results of their toxicological testing pro-grammes in the form of Material Safety Data Sheets (MSDS) for their products. Information contained in the MSDS is regularly forwarded to the national authorities.   The results of epidemiological studies are sent to the National Institute for Occupational Safety and Health (NIOSH), as well as OSHA and the Environmental Protection Agency (EPA).   Of particular importance is the report-ing of toxicologic information to EPA as required by the Toxic Substance Control Act (TSCA).   Receipt of such information from multinational enterprises has prompted OSHA to establish regulations for the material linked to an occupational health problem.   An example of this type of action was the regulation establish-ing occupational exposure for vinyl chloride.

The Netherlands

The changes presently taking place within the Netherlands to implement the Working Environment Act of 1980 offer many possibilities for exchange of occupa-tional safety and health information between the MNEs and national authorities.

The provisions of the Working Environment Act require a substantial increase in the number of norms or regulations which must be established for the guidance of industry, including the operations of the MNEs.

It was noted by the national authorities interviewed that the Shell operating company in the Netherlands had regularly participated on committees to develop occupational exposure limits for chemical substances (MAK values). The national authorities anticipated an increase in the activities of committees developing occupational exposure limits for a larger group of chemical substances and physical agents in the workplace. The committees typically consist of representatives from industry, the workers, the national authorities and experts invited to provide additional research information.

The national authorities noted that Shell had released the results of toxicological studies performed for a variety of Shell products, whereas other MNEs manufacturing products with the same component materials did not release similar information about their products. In one instance, the release of this information resulted in Shell losing some of its market share for one type of lubricating oil found to be moderately carcinogenic, even though the products of competitors contained a higher concentration of the carcinogenic material than the Shell product. The national authorities considered that this form of uneven information disclosure was a problem and were considering regulations to require uniform disclosure under the new Working Environment Act.

Another provision of the Working Environment Act requires an analysis by the enterprises of the country of the potential for a major calamity within the operations, and in turn a plan by the enterprise to prevent such a calamity. Since many of the firms in the refining and petrochemicals industry are MNEs, a significant amount of information will be provided to the national authorities, who in turn must review the information and discuss it with the enterprises before granting acceptance of the analysis.

In the case of the Merck, Sharp and Dohme operations in the Netherlands, information was exchanged with the national authorities regarding some of the special hazards of manufacturing pharmaceuticals, such as that mentioned earlier of explosions of powders during processing operations. Other information was exchanged with the national authorities regarding the occurrence of dermatitis in some workers where the handling of the same material in other countries had not produced similar cases. Special precautions were therefore implemented to prevent the recurrence of this form of dermatitis for the Dutch workers.

The United Kingdom

A unique opportunity exists within the United Kingdom for MNEs and other enterprises to communicate with the national authorities. The Accident Prevention Advisory Unit (APAU) acts as an investigatory and advisory group as implied by its title, within the overall programme of the Health and Safety Executive. The APAU has been performing studies of the safety and health programmes within large enterprises, often multinational, in the United Kingdom. The studies are undertaken with the mutual agreement of the enterprise and the APAU. The reports are confidential to the enterprise and are seen by the Executive. Because of this specialised function, the APAU has had many opportunities to interact with MNEs in the course of performing its studies. The expected outcome of a typical study is an evaluation of the current safety and health performance of the enterprise, accompanied by recommendations for improvements. It is obvious that the conduct of such an extensive study would require an assessment of the safety and health standards used in the workplace, as well as the success of the enterprise in attaining the standards.   The APAU staff has the opportunity to develop a broad background in occupational safety and health prevention measures as a result of viewing many different workplaces across a variety of enterprises, and at the same time to share this information with the enterprises being evaluated. The BICC Group is presently participating in a study by the APAU as noted earlier in this report.

The APAU produced a communication in 1979 which contains a number of observations regarding the safety and health standards of multinationals.   Some of these observations are summarised as follows:

(1)   The safety and health performance of large enterprises operating at multiple sites tends to be variable, despite a uniform policy regarding safety and health within the enterprise.   Some of the reasons for this variability are the age of the equipment within the factory, the geographical location of the site within the country, and the type of industry in which the different sites are engaged.

(2)   If safety and health performance is measured by the number of prosecutions brought by the factory inspectorate for breaches of health and safety legislation, the multinational enterprises operating in the country perform better than average when compared with British firms of comparable size.

(3)   The avowed policy of most multinationals is to follow their own standards unless those of the country in which they are operating are higher, in which case they would be followed.   A general conclusion of the factory inspectorate was that the United Kingdom was one of the few countries where its national standards were higher than those aimed at by the headquarters' unit of foreign-based MNEs.   This was especially true with regard to machinery guarding.

(4)   The safety and health performance of an MNE is very much a reflection of the overall quality and competence of management, i.e. well managed com- panies in terms of market share, profitability, etc., are also likely to have good safety and health records.

(5)   The managers of foreign MNEs often set a high standard of safety and health performance because they are sensitive to the social pressures of being guests within the country as compared to the local managers.

(6)   It was noted that an increasing number of British companies incorporate a statement on safety and health performance as a part of the annual report, and that such a requirement uniformly enforced for all enterprises would help create an overall environment for the improvement of occupational safety and health.

It will be recalled that a major cornerstone of the HSW Act is the require- ment that each enterprise develop a safety and health policy and the measures necessary for the implementation of this policy.   Of course, the policy must be conformance with the national legislation of the United Kingdom. Many oppor- tunities obviously exist for an exchange of information between the factory inspectorate and the MNE in reviewing this document.

Switzerland

There are some similarities in the roles played by the Industrial Mutual Accident Insurance Associations of the Federal Republic of Germany and the Swiss National Insurance Bureau (SUVA), in that each organisation exerts a major influence over the establishment of health and safety regulations in the workplace.   Because the competence of the various cantonal authorities in enforcing occupational safety and health is much more varied than that of the state inspectors in the Federal Republic of Germany, SUVA assumes a very important role in guiding the accident prevention programmes of enterprises throughout Switzerland.   Thus, it is very important that the multinational enterprises operating within the country exchange information with SUVA officials as well as with the cantonal authorities.   Because the practical implementation of the federal ordinances governing occupational safety and health is left to the cantonal authorities, which vary a great deal in their competence, the type and amount of information exchanged between multinatinonal enterprises and the cantonal authorities varies throughout the country.   In a canton such as Geneva, the authorities assume the role of a well-trained factory inspectorate requiring much more detailed information from multinational firms than in a canton where the authorities may have less competence in occupational safety and health matters and in turn require much less information because of their inability to evaluate adequately such information.

Mexico

The MNEs operating in Mexico stated that they had only a limited contact with the federal factory inspectors, whereas they had a larger number of contacts with accident prevention specialists of the Social Security Institute.   The reasons for the limited contact with the federal factory inspectors were explained by the Director-General of the Federal Factory Inspectorate during an interview.   A system is used to aim the visits of federal factory inspectors at those enterprises which have the poorest safety and health records within the country.   With approximately 230 federal inspectors to visit enterprises throughout the country, it is difficult for the inspectors to spend very much time with the enterprises having the better safety and health records which normally include the MNEs because of the time which must be spent in visiting the others.

The Director-General of the Federal Factory Inspectorate noted that a significant opportunity for exchanging occupational safety and health information with MNEs occurred during the meetings of safety and health committees appointed for the major industrial sectors to establish programmes and standards within the industry.   These committees have representatives from the Federal Factory Inspectorate, the states of the Republic, workers, employers, the Social Security Institute and the Ministry of Health.   Information regarding safety and health of the industries within the region is supplied to the committee, which then establishes goals for accident prevention and advises the Federal Factory Inspectorate on the adoption of standards.

The Social Security Institute works in consort with the federal and state factory inspectorates to investigate accidents and provide advice to enterprises regarding accident prevention.   Because of their more frequent contacts with the accident prevention specialists of the Social Security Institute the MNEs participating in this study noted that they had an opportunity for an exchange of views and information regarding injury and illness prevention in the workplace. The MNEs themselves stated that the contacts with the Social Security Institute were useful, especially in the adaptation, where required, of their own safety and health standards to the laws and conditions existing within the country.

Nigeria

The Nigerian authorities tend to view subsidiary operations of multinational enterprises in the country for all purposes, including safety and health matters, as national enterprises, since they are incorporated in Nigeria and the Government owns a majority interest. In the case of Shell and Volkswagen, the MNE ownership was 40 per cent, while the government share was 60 per cent.

The MNE subsidiaries noted that visits by factory inspectors were somewhat rare and, while this was no doubt related to the shortage of manpower resources and transport facilities for the inspectorate, it tends to confirm the fact that the national authorities viewed the safety and health performance of Shell, Volkswagen and Balfour Beatty as being superior to those of most of the entirely domestically owned enterprises. There is thus a tendency for the inspectorate to direct their attention to the latter enterprises. The MNEs were viewed as an important source of information for measures to protect the safety and health of workers. Various ways were available for this exchange of information, including the following means:

(1) The participation of multinationals on committees with workers and the national authorities to provide information concerning standards for the protection of workers' safety and health.

(2) The training of nationals employed within the MNEs to increase the pool of qualified safety and health specialists.

(3) The creation of internship positions within the MNE in order to assist in the training of factory inspectors and health specialists.

(4) The participation of MNE staff members in local accident prevention and related activities.

Communication and co-operation between
MNEs and employer associations

The MNEs participating in the study are members of employer associations in the home and host countries where such associations exist, offering an opportunity for the exchange of safety and health information. However, in the United States the exchange of safety and health information among multinationals and national firms takes place mainly through committees of professional associations and other safety and health interest groups which may be formed among the largest enterprises within a given industry or group of industries. The Central Union of Swiss employers' Associations noted that there are relatively few subsidiaries of foreign multinational enterprises operating in Switzerland, apart from banking and insurance. Therefore, multinational enterprises within the manufacturing sector of Switzerland consist primarily of those with headquarters in the country.

The MNEs gain experience from operating in a number of different countries with different safety and health programmes and can therefore provide helpful information to national employers within the association. However, the training and experience of safety and health staff members in multinational enterprises may not be any greater than that of the safety and health staff in comparable enterprises within the industrialised countries and thus the sharing of information does not have the same importance as it does in the developing countries. Employers' associations in both Nigeria (Nigerian Employers' Consultative Association) and Mexico (CONCAMIN) both referred to the contributions made by multinational employers in the training of national employers regarding occupational

safety and health,   For example, a large MNE had provided instructions for the
handling and use of explosives to a group of employers within CONCAMIN.

Communication and co-operation between
MNEs and workers

Of all the various forms of communication and co-operation in the area of
occupational safety and health, perhaps the most important and direct communication is that which occurs between the MNEs as employers and the workers in the
various units of the enterprise.   This section focuses both on this internal
communication and that which exists between various worker organisations and the
MNEs. Since in most cases, the nature of this communication and co-operation
is shaped by national legislative provisions, the information which follows is
arranged country by country according to the location of the headquarters or subsidiary units of the MNEs participating in the study.

Federal Republic of Germany

The overall systems of industrial relations and occupational safety and
health within the country provide many opportunities for an exchange of information between multinationals and workers, as well as for co-operation in the protection of workers.   At the plant and higher enterprise levels the works council
("Betriebsrat") provides the opportunity with which management discusses safety
and health and other relevant measures within the plant which are of interest
to the workers.   In addition, an occupational safety and health committee
(linked with the works council) must be established in the larger enterprises
and this presents a further opportunity for exchange of viewpoints and co-
operation in protecting workers.   Of course, both MNEs and national enterprises
have to follow the same legal provisions with regard to the operation of the
works council and the safety and health committees.

It was noted during interviews with the principal workers' union organisation the Deutsche Gewerkschaftsbund (DGB) and the Metalworkers' Union (Industriegewerkschaft Metall or IG Metall) that a very well-developed programme exists
within these unions to train their workers in occupational safety and health.
The IG Metall runs a special school to train safety and health instructors who
are assigned to the local branches of the union and operates a very active safety
and health programme for its 3.7 million members, publishing a large amount of
educational material as well as undertaking training in the field of occupational
safety and health.   All of the district branches of the union have safety and
health committees to aid in understanding safety and health issues and to help in
eliminating hazards from their workplaces.   The IG Metall has also given strong
support to activities in the field of safety and health at the conference of
trade unions held every six years.   It was noted that IG Metall maintains close
liaison with the International Metal Workers' Federation which has a world-wide
programme of worker education in safety and health issues.

The DGB also operates a safety and health programme for the benefit of its
member unions.   The general impression of the union officials interviewed was
that the large enterprises, including MNEs, were more advanced in occupational
safety and health matters than the smaller enterprises. In general, the sharing
of occupational safety information by the MNEs was considered to be satisfactory,
where some reserve was expressed about occupational diseases concerning which
it was suggested that more advance information should be voluntarily shared.
The union felt strongly that testing of new substances should be intensified,
and that where health hazards were found or suspected, early action even to the
extent of banning the substance concerned, should be initiated.

Because the national labour legislation grants workers or their representatives rights to discuss safety and health issues with their employers especially within the framework of the Berufsgenossenschaften, and in the case of the works councils to actively promote the improvement of safety and health mesures in the workplace, the collective bargaining agreements between workers and their employers have few provisions regarding safety and health measures. The only measures reported to be included in some collective bargaining agreements concern the employer supplying protective clothing for workers in the enterprise.

United Kingdom

The legal requirement that each enterprise establish a general policy with respect to the health and safety at work of their employees, as well as to the arrangements for implementing the policy, provides an important means for communicating the safety and health programme of the enterprise to the workers. The regulations under the HSW Act may (i.e. according to the size of the company) further require the establishment of a health, safety and welfare committee within the enterprise, with safety representatives appointed by trade unions to represent the workers and to exchange information with their employer regarding safety and health concerns. The workers interviewed at the MNEs within the United Kingdom stated that the means available for the exchange of occupational safety and health information were satisfactory, but they felt that there was a need for more education of workers regarding safety and health issues. In their opinion, the workers of the United Kingdom were not as well educated regarding occupational safety and health matters as some of their counterparts, for instance in the Federal Republic of Germany. It was noted that the relatively larger number of unions within the United Kingdom made it difficult for any single union to mount a large-scale programme of safety and health education.

The BICC Group management referred to the use of "care sheets" which were supplied to workers regarding the hazards of substances found in their workplace and the precautions necessary in working with the material. Worker interest in obtaining such information from their employers has been heightened by educational activities designed to provide information about substances present in the workplace which can produce serious illnesses, especially those which can produce cancers of various types.

In addition to the safety representatives as required by the regulations, the Welwyn Garden City plant of the Rank Xerox subsidiary within the United Kingdom appoints a special official ("safety landlord") to a specific area or section of the plant to facilitate the exchange of information on safety and health matters, to identify safety problems and to provide an easy focus for workers within the area to transmit their concerns to management for a response. This innovative approach was praised by both the enterprise and the workers as an excellent means of exchanging occupational safety and health information.

United States

There are no statutory provisions requiring the appointment of a safety and health committee within an enterprise to facilitate the exchange of occupational safety and health information between the workers and their employer. In general, the management of an enterprise considers that they are competent to deal with matters of safety and health in the workplace and are therefore unwilling to share any decision-making with workers regarding the measures which should be taken for protecting the workers' safety and health. In the absence of national statutes requiring that safety and health committees be established within an enterprise, a number of the major workers' unions within the United States have negotiated collective bargaining agreements to provide for the appointment of a worker-management safety and health committee. In general, the safety and health committees which operate serve principally as a means of exchanging information and setting goals for safety and health performance, without granting workers any voice in the decisions which are made regarding the provisions for safety and health in the workplace. However, as a result of the information gained during the safety and health committee discussions, the workers' unions have negotiated collective bargaining provisions relating to specific safety and health measures in the workplace. In addition to serving as a forum for discussion of safety and health matters, the committees may also conduct safety surveys within the enterprise.

The collective bargaining agreement between the workers' unions and the management of the Wood River, Illinois refinery of the Shell Oil Company contains provisions governing the operation of a worker-management safety and health committee. A similar committee was found to exist at the Xerox facilities visited in Webster, New York. The Bechtel Power Company has established a safety and health committee at one of its major construction sites in the State of California which has attracted a large amount of interest among the management of enteprises within the construction industry as well as other sectors of industry. The safety and health committee was considered unique because it assumed many of the inspection functions of the factory inspectorate (OSHA compliance officers) in responding to worker complaints of unsafe conditions in their workplace as well as inspection of the work site to detect violations of OSHA regulations. The Bechtel committee serves as the focus for exchanging information regarding occupational safety and health as well as the function of arbitrating disputes regarding the application of OSHA regulations in the workplace. The compliance officer visits the site to enforce the law only when the arbitration process proves unsuccessful, which has rarely happened.

The workers' organisations also benefit from the Material Safety Data Sheets (MSDS) referred to earlier in this report. The Shell Oil Company has been in the forefront of enterprises preparing guides for the use of MSDS, and has provided a detailed glossary of the technical terms commonly used. In addition, the Shell Oil Co. Chemical Division has produced a chemical safety guide for all of the products manufactured, giving information regarding the health and safety hazards of the material, procedures to be used in the event of a spillage or leakage of the material, protective clothing to be used when handling the material, and first-aid measures to follow if accidental exposure to the material occurs. Shell management noted that several hundred thousand copies of the two publications had been printed and distributed in reponse to an unexpectedly large number of requests from both workers' organisations and enterprises throughout the country.

The Xerox Corporation has prepared a document which defines its safety and health policy and outlines the various measures taken to assure the safety of all its products, workplaces and employees. This information has been widely distributed to employees and customers for the purpose of reassuring workers using Xerox reprographic machines that the machines and materials are not hazardous in normal use.

The Netherlands

The Working Environment Act of 1980 (in force as of 1 January 1984) establishes the requirement for the enterprise to have one or more safety and health committees to promote a concern for safety, health and welfare in the

establishment.   Requirements are set forth for appointing the worker representatives to the safety and health committees.   The duties of such committees are
to deal with matters related to the safety, health and welfare of the workers in
the establishment and to enter into regular consultation with the employer and
express opinions on such matters.   The employer is required to obtain the con-
sent of the committee regarding any decision he intends to take in connection
with the introduction, alteration or cancellation of arrangements relating to
safety, health and welfare at work.   It is therefore clear that an effective
mechanism will be established for exchange of information regarding occupational
safety and health as well as a basis for co-operation in the development and
implementation of measures to protect the safety and health of the workers.

Switzerland

Safety and health committees established on the basis of collective agreements exist within the enterprises with representatives of the workers appointed
to the committee to serve as a communication link with the employer, as well as
setting goals for the reduction of accidents.   A representative of the Christian
Metal Workers' Union (CMV) noted that Swiss workers were becoming much more concerned about safety and health problems in their work environment and that they
were expressing greater interest in obtaining information regarding these matters
as well as having a say in the measures taken by their employer for the protection of their safety and health.   The representative of another union, the
Federation of Textile, Chemical and Paper Workers (FTCP) discussed the incorporation of references to safety and health matters in collective agreements negotiated for the chemical industry in the Basle region covering mainly the three
major Swiss-based MNEs in this sector (Hoffman-La Roche, Ciba-Geigy and Sandoz).
It was noted that workers in the chemical industry had a need for a better understanding of the safety and health hazards of the materials with which they
worked and the FTCP was currently organising courses in toxicology and occupational medicine for members of the safety committees at different enterprises.
The enterprises provide payment to the workers for the time spent in attending
the courses in some instances, while the union compensates the worker when the
enterprise does not pay.   In general, the FTCP considered that they could obtain
any relevant information concerning occupational safety and health from the
larger chemical companies with multinational operations, but that this was not
true for the smaller national firms operating in the country.

It transpired during interviews with Swiss trade unionists that some of the
Swiss MNEs in the chemical sector (not included in the study) had encountered
problems with health hazards and in one case a complete production process had
required modification to provide greater protection for the workers.   It was
stated, however, that workers in the developing countries continued, in some
cases, to be exposed to toxic chemicals due mainly to the lack of knowledge among
the workers in those countries about the health hazards in question.

In this connection, the question was raised by trade unionists regarding the
provision of information on toxic products to the workers at an earlier stage
than was at present the practice.   The management of some companies interviewed,
however, felt that this would present a number of difficulties both from the
point of view of the workers themselves and from the point of view of the company
in particular with regard to the effect such disclosure of information might have
in regard to their competitors.   (A case was quoted where Shell had withdrawn a
product for this reason, and the market demand had subsequently been filled by
competitive companies.)

Mexico

The national legislation requires a safety and health committee to be appointed in each enteprise which must include representatives of the workers and members of management. These committees must meet once a month to discuss safety and health problems within the enterprise and their solution, and produce a report which is distributed to all its members as well as to the national authorities.   The operation of such committees serves as the principal means for establishing communication and co-operation between workers and management within both multinational and national enterprises in the country.

A representative of a workers' union (CTM) noted that there is a serious lack of information particularly with regard to changing workers' perceptions of occupational hazards and that the CTM was conducting seminars to help supply the necessary information.   It was noted that even though asbestosis and silicosis have long been recognised as occupational diseases, many workers still do not appreciate the hazards of exposure to these materials.   In the opinion of the CTM spokesman, the multinational enterprises in Mexico generally have more knowledge and are more interested in providing this information to the workers than national firms.   However, he felt that multinational enterprises should continue their leadership in this manner and provide even more information for the education of workers in occupational safety and health matters.

The CTM official interviewed said that it was very difficult to negotiate collective agreements concerning occupational safety and health matters where the workers did not appreciate the value of such agreements in comparison with those having direct economic benefits, such as wage increases.

Nigeria

An official of the Nigerian Labour Conference (NLC) noted that the Nigerian workers were not well informed regarding safety and health matters other than the provision of protective clothing such as boots or gloves. He believed that serious deficiencies exist in the workers' knowledge of the health effects resulting from exposure to hazardous substances.   In general, he considered that the multinational enterprises in Nigeria provided a much higher level of worker information regarding safety and health matters than the national firms.

Communication and co-operation between MNEs and international organisations

It was found that most of the MNEs participating in the study were familiar with the activities of international organisations such as the ILO and WHO, having participated in one or more activities of these organisations in the past, although certain of the MNEs had little knowledge of the programmes of the international organisations in the area of occupational safety and health.   A brief analysis of the previous contacts between the MNEs participating in the study and international organisations is presented in the following sections, along with suggestions for possible future participation of the MNEs who have not previously co-operated with the international organisations.

BASF

Representatives of BASF were quite familiar with the programmes of the ILO and the WHO, having already exchanged information regarding occupational safety and health.   They expressed a willingness to continue co-operating with

international organisations in the area of occupational safety and health by
providing information, serving on committees of experts, and in other ways in
which their experience could be utilised.

Merck and Company, Inc.

The managers of Merck and Company who were interviewed were familiar with
the activities of the ILO and WHO in the area of occupational safety and health.
Some staff members from Merck have participated in ILO activities in the past
and expressed a willingness to continue to participate in the future.

Royal Dutch/Shell Group

The parent company as well as the operating companies were familiar with
the activities of the ILO and WHO in the area of occupational safety and health.
It was noted that information had been supplied upon request in the past and
that a member of the subsidiary in the United States had served as a staff
adviser to the Employers' group during the last two meetings of the ILO's
Petroleum Committee.

The Bechtel Companies

While some representatives of the Bechtel Companies were not aware of the
programmes of the ILO or WHO in the area of occupational safety and health,
others were knowledgeable of ILO work in the field of accident prevention.
However, they all expressed interest in obtaining more information about these
activities and expressed a willingness to participate in any future meetings
of the ILO Industrial Committees or meetings of experts.

The BICC Group

BICC Group management was familiar with the health and safety activities of
the ILO and WHO as well as other international and regional organisations. They
had participated in previous activities in these organisations and expressed a
willingness to continue to do so in the future.

Volkswagenwerk AG

Headquarter's representatives of Volkswagen were familiar with the activities of the ILO and WHO in the area of occupational safety and health.
Volkswagen officials in Mexico were not as well acquainted with the activities
of the international organisations, but they expressed a strong interest in
participating in future activities of ILO where they could exchange information
regarding their experience in occupational safety and health with other partici-
pants in the meeting. Volkswagen management in Nigeria were aware of ILO activities, having co-operated in connection with a previous research project. They
would like more information about the ILO's activities in the field of health
and safety.

The Xerox Corporation

Managers of the Xerox Corporation were familiar with the activities of the ILO and WHO, noting that the Conventions and codes of the ILO would be considered in drafting a Xerox safety and health standard. In addition, it was noted that Xerox follows the safety and health developments within the European Economic Community and the activities of other international organisations throughout Europe. They expressed a willingness to participate in any activities of the ILO or WHO involved in exchanging occupational safety and health information.

Brown Boveri and Company, Ltd.

Representatives of Brown Boveri and Company in Switzerland and the Federal Republic of Germany were familiar with the activities of the ILO and expressed a willingness to participate in exchanges of information regarding occupational safety and health. In Mexico the official of the company interviewed was not familiar with the activities of the ILO or other international organisations concerned with occupational safety and health.

MNE assistance to professional associations and educational institutions

In every industrialised country several professional associations exist which are concerned with occupational safety and health, i.e. accident prevention, occupational medicine, safety engineering, occupational hygiene, etc. The professional associations serve useful functions in setting standards of practice and providing a forum for the exchange of information about occupational safety and health.

Most professional staff members interviewed in the MNEs were found to participate actively in their respective professional associations, thus sharing information with colleagues and learning of the latest developments in the field. They had been active as leaders in these associations, and in several cases they had served on special committees or task forces to prepare reports or codes of practice for a problem in the area of occupational safety or health. Being able to draw upon the world-wide experience of the MNE was considered a definite asset by the professional staff members serving in such capacities.

The chief medical officer of one MNE noted that he regularly gave lectures on occupational health in the university of a developing country when he visited the subsidiary there. In other instances the MNEs sponsored research in the educational institutions of industrialised countries and retained researchers at these institutions to act as consultants for their safety and health programmes. An area of special importance for such exchanges was research related to the health effects of chemical hazards.

Safety and health information activities of international workers' organisations

An important part of the present study consisted of interviews with international workers' organisations (mostly those based in Geneva) whose interests included industrial sectors related to the MNEs taking part. The names of the organisations contacted are to be found in Appendix IV.

A particular point that emerged from these discussions was that while those organisations representing workers in industries such as mining, metalworking and chemicals have traditionally been concerned about health and safety problems, the impact of new technologies has now given rise to problems in sectors which previously had not thought of such matters. A typical instance is the case of office, bank and other commercial workers, whose health and safety problems had, until recently, not been the subject of detailed research and about which there was a general lack of information among employees. The point was stressed that where information is lacking, rumours, often unsubstantiated, tend to circulate.

It thus became clear that these organisations were engaged in an important expansion of their activities to include health and safety questions. However, some representatives of the international trade secretariats expressed the view that responsibility for safety and health matters ought to lie squarely with management and should not be a matter for bargaining. Their efforts were being directed basically along two lines - to increase awareness among national trade unions and workers of health and safety matters, and to institute training programmes on this subject, particularly among their associated trade unions in the developing countries.

The nature of the hazards that were referred to included substances used in food-packing industries, chemicals and equipment in the printing and allied trades, noise, stress and fatigue and muscular-skeletal afflictions. Ergonomic problems related to new office machinery were also cited. In many cases, rumours tended to circulate among employees who have only a rudimentary understanding of the machines they operate, giving rise to apprehension and alienation from their working environment.

In most cases good relations were maintained with MNEs and while it was felt that safety and health standards were generally satisfactory among the home-based establishments, a number of the unions considered that this was not always the case with the subsidiaries in the developing countries where safety and health problems were met with indifference. In some instances this was attributed to the weakness of trade unions in these countries and sometimes to inadequate legislation and/or enforcement thereof. Particular mention was made of the need for greater research into the problems mentioned earlier concerning new technology in offices and commercial enterprises, a subject upon which some of the MNEs taking part in the present study are showing considerable interest and concern.

Without exception, the international trade union organisations interviewed considered that as far as the developing countries were concerned, the principal need if accidents were to be avoided and greater protection afforded for the workers was centred in the educational field. This need went further than the simple provision of information, since without sufficient knowledge of the problems involved and the possible consequences, trade union officials, workers, and, in some cases, management, were not sufficiently aware of the problems to be able to put such information to any useful purpose.

Note

[1] ACGIH: Threshold Limit Values for Chemical Substances and Physical Agents in the Work Environment, 1983, published by ACGIH, 6500 Glenway Ave., Bldg. D-5, Cincinnati, Ohio 45211, USA:

CHAPTER IV

SUMMARY AND CONCLUSIONS

This study has investigated the occupational safety and health standards in MNEs and how the pertinent information on this subject is transferred from the parent enterprise to its various entities - taking the international experience of the MNE as a whole into account.

In-depth interviews were conducted with management and workers, followed by plant visits and on-site inspections.   Information was further solicited from the ILO's tripartite constituency.   In this connection, co-operation was examined between the MNEs and the competent national authorities, the employers' and workers' organisations and the relevant international bodies in the field of safety and health.

Although the findings of the study have been derived from a sample of enterprises and situations, they illustrate well the types of issues generally found in the area of occupational safety and health and the related way in which information is provided by multinational enterpises.   It is hoped that the findings of the study will contribute to an informed discussion of the subject.

Regulations and standards regarding safety and health

It was found that the safety and health legislation and practices in the home countries of the MNEs studied constituted the basic framework throughout their operations.   These standards were adapted as necessary in the light of local laws where the subsidiaries are located and, in some cases, expanded or complemented by their own more stringent standards.   The general development over recent years of safety and health laws and standards in industrialised and developing countries reflects a world-wide concern for increased protection of workers and a growing awareness among governments, employers and workers of this subject.

Technological change, the emergence of new products and production processes as well as the growing industrialisation of the developing countries has brought the question of occupational safety and health into a new focus.   These developments are also reflected in a renewed emphasis in recent years on ILO standard-setting in this field, and many countries have made progress in the ratification and/or application in practice of the relevant ILO instruments.[1]   Together with these international instruments are the supporting Codes of Practice and Guides, which were known and appreciated by many of the respondents interviewed for the study.

However, it is still true that safety and health standards in the Third World countries are less developed than in the industrialised countries.   It follows from this that especially in developing countries the MNEs, by virtue of their international experience, can play an important role in the improvement of the occupational safety and health of the workers.

Collective agreements for safety and health

The respondents were asked in every case for their opinions about the incorporation of safety and health matters in collective agreements. Although it was found that this practice was followed in some countries, the majority of those interviewed indicated that the safety and health protection of the workers was entirely a management responsibility, subject to laws and regulations. Therefore, the whole question was not one which ought to be the subject of bargaining and possible labour conflict. It may be added that, where provisions in collective agreements exist, they usually relate to matters such as the issue of protective clothing, or the establishment of safety and health committees where these are not already prescribed by law.

Organisation of safety and health in the enterprises

In setting out the principal findings which can be drawn from the present study, it would appear that no uniform pattern of safety and health organisation emerged which applies to all MNEs. Variations existed along sectoral lines - i.e. the type of industrial activity concerned; along national lines, depending on the home and host country of the enterprises concerned: according to the type of control or financial participation of the parent company in the subsidiary operation; and, finally, according to the individual approaches of the company concerned.

The study of the organisational structures of the safety and health function revealed a wide degree of variation ranging from highly centralised control to no central direction from the headquarters unit. However, all of the MNEs participating in the study stated that they do furnish technical advisory services and information in the field of safety and health to their subsidiaries in the various countries.

Safety and health policies and practices

The inquiries undertaken, especially those in the home-based operations of the participating MNEs, demonstrated that safety and health constitutes an important preoccupation of all these enterprises. This was reflected by the level to which safety and health matters are reported, i.e. at managing board level or equivalent; by the organisation and number of staff who are responsible for safety and health matters; by the degree to which safety and health records are taken into account in assessing mid-managerial performance; by its frequent discussion at shop-floor level between junior management and foremen; by the active involvement of works councils or their equivalent in regard to safety and health; and by the considerable interest in the subject shown by workers, safety stewards or committees and trade union representatives interviewed.

The practice of formulating a written safety and health policy - which is even mandatory in some countries - is followed by most of the MNEs. In many cases, the policy is established at top level and includes detailed instructions applicable in each and every operation and plant throughout the group. In other cases, the main principles of the company's safety and health policy are established at headquarters and relevant details and instructions based on these principles are set out by the different plants with the co-operation and approval

of the central safety department.   The safety policies, with their supporting
instructions where appropriate, are available to all concerned and are backed up
by a competent safety and health staff.   These commit line management to an
unequivocal, although not always flawless implementation of the requirements.
Some trade union representatives noted that for certain operations in developing
countries more training would need to be provided with a view to changing the
workers' perception of occupational hazards.

Considerable differences were noted in the way in which safety and health
information and rules are exchanged between the different units of the MNEs.
These ranged from highly detailed written guide-lines published, as mentioned
earlier, as part of the overall safety policy and applicable in principle
throughout the group, to a loosely-structured arrangement whereby the parent company supplied information upon request.   One company noted that its middle
management, when they were posted to the subsidiaries for a limited number of
years, kept up with the safety and health practices in their parent enterprise
because this knowledge would be required upon their return.   This is another
informal way in which standard safety and health practices may become standardised in the various company operations.

The organisation and implementation of safety and health systems and practice was found to be more centralised and developed in the case of MNEs in high-
risk industries, particularly in the chemical and the construction industries.
Most of the MNEs visited were found to make use of centralised computer information systems and accident reporting as well as analysis and safety audits to
improve safety and health standards throughout the different plants.   All
carried out research (usually located in the parent company) regarding accidents
and health hazards from exposure to new machinery or products.   It became clear
in the study that the MNEs falling into this category were the ones with the
most extensive and integrated corporate systems for implementing safety and
health policies and the greater degree of centralised control.

While positive results were observed in all the enterprises, whatever
organisational pattern was chosen for the organisation of safety and health
measures, it would appear that those structures which call for active headquarters involvement both in a guiding and a directing role offer particular
advantages.   Undoubtedly, the more the safety and health function is seen as
a central company responsibility, the more effectively relevant information on
specific and new hazards can be researched and related protective standards be
applied and transferred to all subsidiary operations.   This seems important,
especially with regard to subsidiaries in some developing countries where the
national safety and health standards do not entirely reflect the requirements
for certain new production processes characteristic of MNE operations.

Without exception, the participating MNEs appeared to fully comply with the
safety and health standards required by the national authorities in the home and
host countries studied.   In nearly all cases, the representatives of the national
authorities stated that MNEs' health and safety standards went beyond these
requirements.

In the industrialised countries the implementation of local standards was
normally better for multinationals and large enterprises than for smaller enterprises.   In the developing countries, the MNEs were considered by most of the
respondents to be in the top category of enterprises or, occasionally, in the
medium category judged by available safety and health statistics, although the
difficulty of making statistical comparisons should not be overlooked.   These
findings are more generally confirmed by the results of a larger ILO survey
undertaken in connection with the follow-up to the Tripartite Declaration of
Principles concerning Multinational Enterprises and Social Policy for 1980,
1981 and 1982.[2]

All the enterprises included in the study have been making substantial and successful efforts in recent years to reduce safety and health risks and accident rates, especially with respect to their home country operations. This is particularly pronounced in those enterprises involved in high risk sectors.

In comparing the health and safety performance of home-based MNEs with that of the subsidiaries, it could be generally said, that the home country operations were better than those of the subsidiaries in the developing countries. As far as the subsidiaries in industrialised countries are concerned, some cases were found where their safety performances exceeded that of the home country operation. This might be attributed to various factors such as the different statutory frameworks, product variations, type of plant and worker attitudes, etc.

Variations between the performance of operations in the MNE units in the industrialised countries and in the developing countries were often attributed by respondents to the different climatic conditions and the different nature and cultures of the workforce. In particular, the lack of general safety and health training consciousness and awareness was found to constitute a particular risk factor in certain developing country operations. This situation persists in spite of the particular efforts enterprises had obviously made in this regard. This indicates clearly the need for MNEs to see their safety and health role in developing countries as one of continued and long-term education and co-operation with the relevant local institutions.

A particular safety and health problem was found in developing countries with respect to subcontractors working for MNEs. Some of the MNEs studied established contractual obligations to ensure that subcontractors follow the MNEs own safety and health rules. It would be desirable if this practice could be extended. In some countries, legal safety and health liabilities for general contractors are in existence. In other cases local contractors do not take such safety and health requirements seriously, in particular in the case of short-term contracts. These may be of several days duration only and therefore the operations of the contractors are not closely scrutinised.

Special hazards, products and production processes

The study has confirmed that MNEs, being predominantly found in modern, high techology industries, encounter special or new hazards more frequently than the average enterprise. These are connected with both products and production processes, such as new risks associated with the introduction of advanced mechanical aids or of robotisation, and health hazards resulting from the introduction of new materials.

The MNEs studied were aware of these difficulties and have undertaken extensive research into the related safety and health problems. This is especially true in the case of chemical hazards where epidemiological and toxicological research by the MNEs have made a particular contribution to the available knowledge about these subjects.

New hazards have also come to light in activities where they were hitherto unrecognised as exemplified by the rapidly increasing use of video display units and different types of office equipment and processes which are now affecting workers not only in banking and insurance but in all modern office activities. One of the MNEs producing office equipment contributes to substantial research into the related safety and health problems.

While some enterprises stressed that they share newly-acquired safety and health information in various ways with other enterprises and local industries and organisations, or, in certain instances, make public the results of their testing programmes, references were made to situations where there was some hesitation in this respect for reasons of competitiveness. In one case, negative economic repercussions in the form of a reduced share of a given market was reported as a result of having provided certain information. The question of how to overcome such difficulties in the interests of the protection of the workforce requires further study.

Industrial accidents

The widespread concern that has been expressed regarding new risks and health hazards ought not to overshadow the traditional causes of injury and disability which continue to take the greater toll in all branches of industry.

This was seen to be equally true for the MNEs studied and the statistics showed that the most common causes of lost-time injuries to workers included slips, trips and and falls from a height; lifting and handling of objects; the operation and maintenance of machinery; and, particularly in the case of the developing countries, accidents with moving vehicles including motor transport, fork-lift trucks, conveyors and the like.

It appears that those enterprises which establish a written safety policy at top level, backed by a competent safety and health staff and to which all levels of management and supervisory staff are committed, are the ones which can report the greatest success in reducing accidents. It was evident during visits to the industrial sites that the companies which operate a centralised safety system, with regular reporting back to the corporate safety organisation were the ones which best illustrated an overall implementation of accident prevention measures and where a high degree of safety motivation among the workers was most apparent.

Trade union representatives as well as shop-floor workers interviewed in the industrialised countries generally confirmed the efficacy of this approach, but the same could not be said of all the subsidiary enterprises visited in the developing countries. The principal criticism was directed at shortcomings in the field of basic safety and health education and instruction which, in the opinion of many trade unionists as well as management in general, could form the object of even greater attention on the part of some of the companies concerned.

One example of this problem is the use of protective clothing and safety equipment. Although supplied to the workers, difficulties are experienced in ensuring they are correctly used, or even used at all. This is also a matter of better supervision as well as of education. It became clear in discussions with trade union representatives that the needs for basic safety and health education were far from being fulfilled in the developing countries. As regards their own efforts in this field it was noted that the majority of low-paid workers saw the trade union mainly as an instrument for obtaining wage increases rather than additional safety or health measures.

The whole question of safety education is one to which not only the enterprises but unions and governments as well, need to direct more efforts. During discussions with the international trade union secretariats, it was made clear that they are already devoting considerable resources to the education and training of workers in safety and health. The visits made to developing countries, however, showed that further educational efforts are necessary before accident prevention and the protection of health at work becomes a generally accepted principle among all the workers as well as among many members of supervisory personnel. The old saying remains true, that safety at work depends on "Engineering, Enforcement and Education".

Labour inspection

Interviews with the authorities indicated that relations between MNEs and the factory inspectorate were usually good. However, both in the industrialised and the developing countries it was stated that visits by inspectors to MNEs tended to be a rare occurrence. This was attributed generally to the fact that inspectorates normally concentrate their efforts on those plants where the accident record has not been good or where there is a need for conditions to be improved. Reference was also made to the lack of manpower and other resources in the case of developing countries.

Statistical information on safety and health performance

The statistical records on occupational health and safety that were available from the enterprises and countries for the purpose of the inquiry varied greatly in form and in degree of detail. For this reason any direct comparison between one company and another and one location and another should be treated with caution. The reasons for these variations are many. Firstly, an enterprise is under the obligation to prepare its basic statistics in accordance with the requirements of the national authority wherever it operates. Secondly, although many enterprises enter into greater detail than is required by the competent authority, their methods of computation and the nature of the date collected may vary from one firm and sometimes even from one subsidiary to another. Thirdly, many of the long-established enterprises follow a system that was introduced long ago producing many years of records, while the newer enterprises - including their subsidiaries in the foreign countries - have not yet accumulated enough data to establish valid statistical measures over time.

A number of MNEs have set up a centralised system for the collection of statistics and they measure their performance both against that of previous years and against that of other subsidiaries of the same group. By and large, such comparisons were found to be more meaningful than those obtained by comparing one MNE with another.

While international institutions, including the ILO and ISSA, provide leadership in the establishment of more uniform safety and health statistics throughout the world there seems to be an urgent need for many MNEs to follow the example of those which have established a standardised system of data collection and performance measurement through the whole multinational group.

Transfer of information on safety and health to workers

During the study it was evident that there is a growing concern about the health hazards related to the many chemical and physical agents which are becoming commonplace in many industries and technological processes. It is not only the chemical industries where this problem is encountered, since such substances

and agents are being increasingly used throughout the manufacturing industry as
well as in activities such as construction and agriculture, and the hazards they
represent extend as well to those concerned with their transport and handling.
Attention was recently drawn to the increasing number of potential hazards to
which workers are becoming vulnerable and it has been stated that of some
45,000 to 50,000 chemical substances available on the market, international
standards for safe levels of exposure have only been established up to the
present time for less than 2,000.[3]

Representatives of the workers with whom this matter was discussed expressed
particular concern about the safety and health risks of the new, still not fully
researched substances and production processes. The view was advanced that in
cases where newly-discovered health hazards have become apparent, the workers
did not receive enough information at a sufficiently early stage to permit recognition of possible risks and the introduction of adequate protective measures.
Some of the trade unions interviewed suggested that not only the known toxic
agents but also suspect substances should be banned rather than attempting to
develop a means of protection. A further point that was raised related to cases
of illness which might have stemmed from an occupational exposure and concerning
which it was felt that workers' representatives should be given more information.

From the point of view of the MNEs, it became clear in the course of the
study that they were attempting to understand the problems connected with new
substances. The management of various companies drew attention to the time-
consuming and expensive nature of the testing and research required to establish
the degree of risk associated with products when workers are exposed at various
levels over time. The MNEs concerned have the necessary resources and the
facilities to investigate these hazards, and do, in fact, play a leading role in
the matter. In a number of instances, the workers' representatives were of the
opinion that the MNEs were more forthcoming in regard to the provision of information on these matters than some national firms. Some of the MNEs visited
have developed a scheme whereby every substance or process used in the enterprise was the subject of a special information sheet which described the hazard,
together with the precautions to be taken, and any particular instruction relating to its use. In some countries the national authorities require that such
information be provided to workers.

The conclusion may be drawn that the MNEs should remain fully aware of the
workers' concern relating to the transfer of information about new substances
and agents and should take all possible steps to alleviate this concern. In
this connection substances whose long-term effects on health are not known and
which may be suspect should be treated with particular care and every effort
made to furnish as much early information as possible and provide protective
measures even if these measures later prove to be unnecessary. On the other
hand, the point was raised by workers' representatives that the introduction of
protective measures without sufficient information may give rise to rumours
among the workers regarding the possible hazards involved resulting in a situation of unease.

A number of international and regional organisations are playing a role in
regard to the timely sharing of information on possible health hazards of doubt-
ful substances or processes. To this end the ILO has introduced a "hazard
alert" system through which early warning is received concerning a suspect substance and such information is transmitted to interested parties, including
enterprises, throughout the world. The MNEs interviewed generally support such
programmes and, where appropriate, take part in these activities and make use of
the information received.

An important aspect of the problem relates to the role of governments which, in many cases, have enacted legislation to require the communication of safety and health information to workers. Even where such statutory obligations do not exist, the study showed that such communication systems were established either through collective bargaining or by other means. Safety and health committees composed of management and workers, together with the necessary specialist advice depending on the hazard(s) concerned, were found to be particularly effective in this connection. It was also noticed in the course of the study that some MNEs have introduced new and more informal methods of ensuring better communication between the shop-floor and management by improving accessibility to higher level managers and thus obtaining early attention for problems. The importance of this informal exchange of information was underlined by comments made in the course of the study to the effect that although an enterprise may have a commitment to a free and frank exchange of information as part of its overall safety and health policy, examples existed where the message from the top was not always effectively passed down the line to the supervisors and the shop-floor itself.

Communication between MNEs and national authorities

The study showed that communications on safety and health matters between the MNEs and the national authorities seemed to exist on a satisfactory basis at the level of the inspectorates and ideas and information appeared to be fully exchanged. However, from the information gathered, it may be noted that the MNEs as a whole have, by the nature of their activities, acquired a broad knowledge of safety and health legislation across the world and could no doubt furnish more valuable assistance to national authorities. There is some doubt as to whether this source of potential assistance is being exploited to its fullest extent, particularly in the case of the developing countries. Based on the evidence found in the countries included in the survey it would appear that governments could often make greater or more specific use of the international experience of the enterprises in the interest of improvements of safety and health standards in the different countries.

Co-operation by MNEs with international organisations

In reply to questions on this subject, it became clear during the study that most of the MNEs, particularly those engaged in the chemical and petro-chemical industries, were aware of the activities of international organisations in the field of safety and health, notably the ILO and WHO. It appeared though that they were less well-known in the United States than in Europe. The MNEs generally were familiar with ILO Conventions and Recommendations, but in a number of cases, especially in the developing countries, the codes of practice were not so well known. Certain companies expressed a wish to obtain more information from the ILO in respect of its safety and health activities. In some of the MNE headquarters, it was made clear that their specialists in relevant fields had taken part in ILO and WHO meetings and were in regular contact with regard to certain occupational health problems.

Co-operation by MNEs in national
safety and health training

Discussions with public authorities, trade union and employers' organisations in the host countries indicate that the MNEs or their representatives provide a safety and health contribution by assisting national centres, universi-
ties, professional associations and other bodies at seminars, training courses
and similar activities taking place.   These contributions are sometimes provided in industrial or educational sectors other than those in which the MNE
itself may operate.   In some cases the MNEs co-operate in national safety and
health training efforts as part of their overall activities in the host country,
but in others it was noted that participation of this nature was left to the
initiative of individual staff members.   A general recognition by all MNEs of
such activities as being part of their corporate function in the developing
countries would be a desirable aim which could do much to enhance the development of improved worker health and safety.

Notes

[1] See in this connection ILO: Summary of government reports on the effect
given to the Tripartite Declaration of Principles concerning Multinational
Enterprises and Social Policy in 1980, 1981 and 1982, GB.114/MNE/1/1/D.1, chapter
on "Safety and Health", pp. 101-105.

[2] ibid.

[3] ILO Information, Director-General's speech, Ottawa, Sep. 1983.

Appendix I

Indicative Interview Schedule

ILO Study on Safety and Health Standards of
MNEs in their Home and Host Country Operations

(with special emphasis on the transfer of relevant
information to their workers, to authorities
and to workers' and employers' organisations)

A.   Safety and health problems, standards
     and protective measures

     1.   What are the government regulations and standards regarding safety and
health for the protection of employees in the country(ies) in which the multinational enterprise operates?   To what extent are basic ILO instruments rati-
fied and/or reflected in such regulations and standards (such as Convention
No. 119 and Recommendation No. 118 concerning the Guarding of Machinery (1963),
Convention No. 115 and Recommendation No. 114 concerning the Protection of
Workers against Ionising Radiations (1960); Convention No. 136 and Recommendation No. 144 concerning Protection against Hazards of Poisoning arising from
Benzene (1971) and Convention No. 139 and Recommendation No. 147 concerning Prevention and Control of Occupational Hazards caused by Carcinogenic Substances
and Agents (1974)?   To what extent are they actually applied by both multinational and national enterprises.   Do the local standards which are relevant
to your enterprise provide for adequate safety and health protection?

     2.   To what extent are the Codes of Practice and Guides in the current
list of ILO publications on Occupational Safety and Health taken into acocunt
in the country(ies) of MNE operation?

     3.   How do the standards of safety and health protection applied by your
enterprise (home country operations, subsidiary operation or enterprise as a
whole) compare with the national requirements, similar national enterprises and
those applied by your enterprise in other countries?

     4.   What are the special hazards, products, production processes, etc.
(if any) existing in your enterprise's operation which require protective
measures?   Do these, or do some of them, constitute new safety and health risks
in the country(ies) and for the industries in which your enterprise operates?
What is the relative importance of old and new risks (examples)?

     5.   In which way, and to what extent, is relevant experience in the multi-
national enterprise as a whole (i.e. obtained from the entities in all countries
of operation) reflected in the safety and health protection adopted for the
various entities (question to home country/headquarters enterprise) or for the
subsidiary interviewed in the host country?

     6.   More specifically, what is the policy of your enterprise regarding the
sharing of knowledge on safety and health standards (including those connected
with new products and processes) among the entities operating in the various
countries, including developing host countries?

7. What is the role of the headquarters (main enterprise in the home country) as regards the examination of the causes of such industrial safety and health hazards and the transfer of knowledge on hazards and protective measures to other entities in the various countries of operation? What is the role of the entities in the various countries/regions of operation? Who is basically responsible for the development of protective measures: headquarters/individual entities?

8. Can you provide recent examples of such co-operation among entities (headquarters and subsidiaries) with respect to specific products/processes/new hazards?

B. Informing workers and co-operation
with authorities, workers' and employers'
organisations

9. In which way are workers and their representatives in the enterprise informed about safety and health problems, standards and protective measures? How are they informed about safety and health standards, relevant to local operations, which the enterprise observes in other countries, in particular as regards special hazards and related protective measures associated with new products and processes? In this connection what role do workers' representatives/organisations play in the development of supervision of protective devices and their observance (as compared to the role played by the enterprise or public authorities)?

10. Have competent authorities and/or workers' and employers' organisations in the country(ies) of operation requested information on safety and health standards, relevant to the local operations, which the enterprise observes in other countries, in particular as regards special hazards and related protective measures associated with products and processes? If this applies, how has the enterprise met such requests? Can you indicate any recent examples of such provision of information? What use has been made of this information by the recipients?

11. What role has the enterprise played in the advancement of knowledge about occupational hazards, standards and protective measures in the country(ies), industry(ies) in question? What role have local organisations played?

12. To what extent, and in what way, are workers and their representatives (and possibly authorities and/or workers' and employers' organisations) associated with the examination of hazards and the development of safety and health standards/protective measures found in the enterprise?

13. To what extent have matters relating to safety and health been incorporated in agreements with the representatives of the workers and their organisations? Where is this appropriate (inappropriate)?

C. General co-operation with national
safety and health authorities and
international organisations

14. Does the enterprise co-operate with the competent safety and health authorities and/or establish safety and health organisations in the country(ies) of operation as regards safety and health matters (identification of hazards,

development of standards and protective measures for country(ies), industry(ies), enterprises, products, processes)?   What form of co-operation exists in this respect (please describe in detail).   Is such co-operation a national practice, i.e. is it usual both for national and multinational enterprises?

15. Does the enterprise co-operate in the work of international organisa-tions (in particular the ILO) concerned with the preparation and adoption of international safety and health standards?   Where it exists, what does such co-operation look like in detail?

D.   Other information

16. Any other information which the respondents may wish to provide with-in the context of the study.

April 1982

Appendix II

Positions of persons providing information at
each multinational head office and site visited

BASF AG

Ludwigshafen (Federal Republic of Germany):

> Legal counsel
> Safety chemist
> Corporate medical officer
> Physician
> Co-ordinator - Latin America
> Member of works council and safety committee

Santa Clara (Mexico):

> Technical director, BASF Mexicana
> Medical service physician
> Safety service engineer
> Supervisor
> Worker representative - Confederacion de Trabajores de Campensinos (CTC)
> Manager of division operations
> Manager of administration

Merck and Company, Inc.

Rahway, New Jersey (United States):

> Director - corporate safety and industrial hygiene occupational safety
> and health
> Manager - safety and industrial hygiene, Merck Sharp and Dohme, Inter-
> national Division
> Associate director of manufacturing services, International Division,
> Merck Sharp and Dohme

West Point, Pennsylvania (United States):

> Manager, safety and industrial hygiene for Merck Sharp and Dohme division
> Chief medical officer for site
> Supervisor in one area of manufacturing plant
> Workers in production area, member of Oil, Chemical and Atomic Workers'
> Union (OCAW), safety representative
> Maintenance worker and member of OCAW, safety representative

Haarlem (Netherlands):

> Operations manager
> Plant engineer
> Safety manager
> Plant physician
> Superintendent, animal health division
> An operator and a mechanic (union members)

Royal Dutch/Shell Group

The Hague (Netherlands):

   Group safety and environmental conservation adviser
   Safety and environmental conservation liaison
   Safety adviser, exploration and production operations

Rotterdam - Pernis Works (Netherlands):

   Plant manager
   Assistant manager of insecticide plant, member of staff union
   Worker in isoprene monomer production, member of Federatie Nederlandse
      Vokbewejig (FNV) union
   Maintenance operations supervisor
   Plant section head
   Manager of instruction and employee information
   Safety and environmental conservation liaison

Houston, Texas (United States):

   Corporate manager of safety and industrial hygiene
   Safety technologist
   Senior safety representative
   Manager, regulation and safety services
   Safety division - Shell development
   Safety - marketing and distribution
   Safety - Shell pipeline company
   Safety - operations
   Safety - operations
   Safety - exploration and production

Wood River, Illinois Refinery (United States):

   Manager, safety and industrial hygiene
   Industrial hygienist
   Worker, member of Operating Engineers Union
   Worker, member of Boilermakers Union
   Maintenance supervisor

Lagos and vicinity (Nigeria):

   Administrative manager
   Safety co-ordinator
   Chief medical officer
   Occupational health physician
   Four representatives of NUPENG, the workers union were interviewed as well
      as representatives of PENGASSAN, the senior staff association

The Bechtel Companies

San Francisco, California (United States):

   Manager of safety and health for the corporation
   Manager of international safety operations

Palo Verde, Arizona (United States):

Manager, domestic safety and health
Occupational physician under contract to Bechtel at construction site
Site safety manager
General foreman, iron workers
Overall site manager
Union steward for Pipefitters Union AFL/CIO

Hammersmith, London (United Kingdom):

Regional safety manager, international operations

Fawley (near Southampton) (United Kingdom):

Site manager
Safety supervisor
Safety representative
Safety representative

The BICC Group

London (United Kingdom):

Executive deputy-chairman responsible for health and safety policy

Prescott Works (United Kingdom):

Operational head for all safety and health activities
Worker representatives

Leigh Works (United Kingdom):

Works safety officer and group safety co-ordinator
Worker representatives

Jos (Nigeria):

Group medical officer
Managing director
Chief of administrative division
Shendaw site agent
Site engineer for Gleesons company
MacAlpine company representative
Lilleys (construction engineers)
Lafia/Obi site agent
Talks with foremen, site operatives and workers

Volkswagenwerk AG

Wolfsburg (Federal Republic of Germany):

Corporate plant - engineering management
Legal department - foreign holdings
Medical officer

Foreign production
Chief safety engineer
Safety engineer
Works council member on safety committee

Puebla (Mexico):

Director of plant engineering
Executive council member
Manager, planning and operations
Manager, safety and fire department
Medical director of plant
Manager, personnel department
Member of board for personnel

Lagos (Nigeria):

Divisional manager of production
Safety engineer
Divisional manager for legal matters
Senior staff association and junior staff union

Xerox Corporation

Webster, New York (United States):

Director of corporate environmental health
Manager of operations safety
Manager of environmental health
Manager of environmental health and safety area
Safety specialist for a production unit
Foreman of toner brush manufacturing department
Worker in toner blending and filling department, member of Amalgamated
    Clothing and Textile Workers Union
Engineering manager for photo receptors manufacturing

Mexico City (Mexico):

Director of planning
Co-ordinator of safety and security

Puente de Vigas (Mexico):

Personnel manager for plant
Worker in plant, member of safety and health committee
Manager of machine reconditioning plant

Aylesbury (United Kingdom):

Manager, occupational safety, health and environmental control

Welwyn Garden City (United Kingdom):

Occupational health nurses
Safety representatives

Brown Boveri and Company, Ltd.

Baden (Switzerland):

    Central safety department
    Works council member of safety committee

Mannheim (Federal Republic of Germany):

    Manager, corporate safety department
    Plant manager
    Chairman, combines works council (and also chairman of Berufsgenossen-
      schaften)

Tlalnepantla (Mexico):

    Manager of the plant

Appendix III

Trade unions, employer associations and public authorities providing information in each country

Federal Republic of Germany

Trade unions:

Deutscher Gewerkschaftsbund (DGB)
Industriegewerkschaft Metall (IG Metall)

Employer association:

Bundesvereinigung der Deutschen Arbeitgeberverbände (BDA)

Public authorities:

Bundesministerium für Arbeit und Sozialordnung

United Kingdom

Trade unions:

Trades Union Congress (TUC)

Employer association:

Confederation of British Industry

Public authorities:

Director of the Accident Prevention Advisory Unit (APAU) of the Health and Safety Executive

United States

Trade unions:

AFL-CIO

Employer association:

US Council for International Business

Public authorities:

Assistant Secretary of Labor - Occupational Safety and Health Administration
Director of Field Coordination (OSHA)

Switzerland

Trade unions:

> Secretary, Christian Metalworkers' Union (CMV)
> Federation of Textile, Chemical and Paper Workers (FTCP)

Employer association:

> Secretary of the Central Union of Swiss Employers' Associations

Public authorities:

> Federal Labour Inspector-in-Chief, First Division (Lausanne)
> Chief, Occupational Health Division of the Federal Inspection Services

National insurance association:

> Director of Accident Prevention Division of the Swiss Accident Assurance
> Bureau (SUVA)

Mexico

Trade unions:

> Secretary of external relations for CTM.

Employer association:

> Confederación de Cámaras Industriales de los Estados Unidos Mexicanos
> (CONCAMIN)

Public authorities:

> Secretaría del Trabajo y Previsión Social
>
> - Director-General of International Affairs
> - Director of Federal Inspection of Labour
> - Sub-director of Technical Affairs
> - Subdirector de Asesoría y Consulta de la Dirección General de
>   Medicina y Seguridad en el Trabajo
> - Director, Instituto Nacional de Administración del Trabajo
> - Chief, Servicos de Seguridad en el Trabajo del IMSS (Social Security
>   Institute)

Nigeria

Trade unions:

> Assistant Secretary General (Education) of Nigeria Labour Congress (NLC)

Employer association:

> Director of Nigerian Employers Consultative Association (NECA)

Public authorities:

    Assistant Director, International Division, Ministry of Employment, Labour
      and Productivity (MELP)
    Assistant Director, Factory Inspectorate, MELP

Netherlands

Trade unions:

    Federative Nederlandse Vokbewegig (FNV), safety and health expert
    International affairs - Netherlands Trade Union Confederation

Employer associations:

    Federation of Netherlands Industry (VNO)

Public authorities:

    Deputy Chief for Chemical Hygiene, Directorate-General for Labour Safety

Appendix IV

International unions providing information

International Federation of Commercial, Clerical and Technical Employees
(FIET), Geneva:

Secretary for banking and insurance sections
Secretary for industry, commerce and related sectors

International Textile, Garment and Leather Workers' Federation, Brussels:

General Secretary

International Metalworkers Federation (IMF), Geneva:

Safety and Health Department

International Union of Food and Allied Workers' Associations, Geneva:

Education Secretary

International Federation of Building and Wood Workers, Geneva:

Assistant to the General Secretary